Afraid of Less

What To Do About the Future

Also By Thomas Corcoran

Novels
<u>Novels</u>
Waiting for V-J Day
Decoration Day
(In Commodore Hilty, Afloat and Ashore):
The Flag List
Commodore Hilty's Second Act

<u>Stories</u>
The Hobbledehoy

Afraid of Less

What To Do About the Future

by

Thomas Corcoran

Bernice Feigenbaum & Company
Philadelphia ~ 2023

Cover image courtesy of NASA, then replicated

ISBN (HC): 979-8-218-26944-9
ISBN (PB): 979-8-218-26945-6

First printing

For my children:
Heather Corcoran
Matt Corcoran
John Corcoran
Patrick Corcoran

*And for all the children
of this century*

Do what you can, with what you have, where you are.

—Theodore Roosevelt, 1913

(1913 CO_2 global emissions: 1 gigaton; 2019 CO_2 global emissions: 38 gigatons)[1]

Contents

Foreword

I wrote this essay during the Covid-19 pandemic. After finishing it in 2020, I put it on the shelf, fearing how it might be received.

Since then greenhouse gases have continued to pour into the Earth's atmosphere, the planet has grown so much hotter that feedback loops have been triggered, and signs of the calamitous future can be seen in almost every region.[2]

The science is ever more alarming. Nevertheless, many people, mostly in the industrialized nations, have done little to reduce their carbon footprints. Perhaps they are still hoping for the best. Perhaps they think that their own actions cannot make a difference to such an enormous problem. Perhaps they have already decided that the calamity is inevitable and therefore they will enjoy their lives as much as possible in the time remaining; some of them may be quietly preparing a sanctuary for themselves and their descendants.

I have no reason but to think that these are good people, responsible citizens, loving parents.

As I review the latest science and see, as we all do, the results of the conditions it describes, I have found that what I wrote in 2020 continues to be true and important. Almost certainly human civilization will not achieve net zero greenhouse gas emissions by 2050; almost certainly the Earth's average temperature will rise more than 1.5 degrees above pre-industrial levels. It's a phenomenon that collapse happens faster than expected. What has taken place since 2020 might be the start of the greatest collapse of all time.

Forward

That being so, is my essay not still useful, perhaps to generations yet unborn? I believe it is. More than that, I believe that I have a duty to offer my thoughts on how we humans caused this calamity and what we may still do to mitigate it—what, I believe, we must do for the sake of our grandchildren and great grandchildren. I publish now in a contentious age quick to condemn unpopular points of view, hoping that any offense caused will be forgiven if my words prove prescient and useful.

What follows, then, is the 2020 essay nearly verbatim. I have corrected obvious errors, updated the argument for major developments, and reviewed it again for accuracy, fairness, and, where appropriate, sympathy. Of course I am responsible for what I have written. Many of my views are radical—easier said than done. Keeping that in mind, I would ask the reader not to ascribe to me or to assume that I hold any views other than those I have stated here.

Thomas Corcoran
Philadelphia, 2023

Introduction

Consider the frustration of environmentalists. For more than thirty years they have been warning about climate change. Not only have we as a society failed to heed their warnings, during those years we have released into the atmosphere half the carbon dioxide of the entire Industrial Age.[3] If we had acted more promptly, perhaps no great sacrifice would have been required of us. Now, only a drastic reduction in carbon dioxide and other greenhouse gases,[4] likely attended by years of hard times, can slow the onset of rising seas, cataclysmic storms, massive wildfires, killing heat, ruinous drought, and recurrent plague—slow but not stop it. If they sound a little strident now, who can blame them?

And there's this: all the industrial nations of the world have contributed to the problem, but in the United States some oil and gas executives were actual villains; for they knew in the 1970s—years before the first public warning[5]—what the burning of fossil fuel was doing to the Earth, yet they lied about it to preserve their fabulous profits. In this they were aided by certain politicians and economists, disciples of unrestrained growth, who were themselves beguiled by the doctrinal novels of Ayn Rand, a self-styled philosopher of an illusory way of life.[6]

In one sense accountability isn't helpful now. The collision between humans and the natural world, as a group of eminent scientists described the problem in 1992,[7] is already taking place. When you are thrown out of your bunk and water pours into the hull and fire fills

Introduction

the passageways is not the time to punish the helms-
man. You have a ship to save.

But that means knowing the damage and understanding what caused it. Toward that end, the emphasis
by some environmentalists on our present failings mis-
reads the record. If half the greenhouse emissions occurred in the last thirty years, after the first public
warning was given, then the other half occurred before
then. And if the world might have done something to
heed an earlier warning, then it might not have, either.
By now the danger is clear, yet few of the major industrialized nations, all of whom signed the Paris Agreement of 2016, are on track to reduce their emissions as
they promised.[8] This includes Canada,[9] whose prime
minister has justified the expansion of mining the Alberta tar sands by stating that no country would leave
so much oil in the ground.[10]

We cannot say with confidence how much would have
been left in the ground if the oil and gas executives had
told the truth when the truth was first known to them.
It wasn't Exxon or Ayn Rand either that caused this collision.

It was us, many generations of supposedly enlightened people; all of us who took more resources from the
Earth than we needed. You and I and our parents and
grandparents and anonymous ancestors going back to
about 1850, with aspirations going back much further
than that. Human history can be written as a study of
waste. In particular the advanced societies of the world
have been on a resource binge since the first steam engine. Even now, under the threat of extinction, support
for growth of all kinds is widespread and powerful. The
world as we know it was built on ever seeking *more.* In
the future we will have to live with *less.*

The danger poses special challenges. First, while
many things can be done, many others cannot, and
wishful thinking will squander our efforts. Second, the
injustice done to those left out of the binge may be re-

Introduction

peated on an existential scale when facing its consequences: the poor will have it worse. And third, the
changes we need to make are more complicated and go
further and deeper than we understand.

In that sense, the problem will be even more difficult
to manage than the direst predictions of the environmentalists. Having built our civilization one way, we
must now rebuild it into something very different; something smaller in impact to support billions more consumers. The original endeavor, which required genius,
fidelity, sacrifice, and courage, rewarded its creators
with personal satisfaction as well as riches. The task
before us now will demand our finest qualities without
the accustomed rewards.

In this Introduction, then, you have the entire book.
Since I am not a scientist, I won't repeat the science except to comment on it. In general I believe that all of us
know the truth, intuitively if not empirically: we just
have to act. And that is part of the problem to be solved,
a discipline in which I do have experience, both defining
problems and solving them.

But none like this....

A Problem As Big
As the Planet

The Age of (...)

First, we need a name—not for this book, we'll come to that, but for the age we've entered. An official name has been proposed and may be approved before these words see print. The Anthropocene Epoch: probably you've heard of it. It's from the Greek for "human" and "new," meaning something we've done recently. An epoch is a unit of geologic time shorter than a period but longer than an age. The current period is the Quaternary, and the current epoch, which the Anthropocene may succeed, is the Holocene, effectively describing the time of humans until now.[11]

Anthropocene sounds scientific and vaguely penitent, so let the International Commission on Stratigraphy, whose job it is, choose that, the sooner the better given the circumstances. Unfortunately, while the term may describe the state of things from Earth's perspective, it doesn't do justice to our own, this time of fear, hesitation, and discord we cannot see the end of. From what I know, it feels like a second Middle Ages, with the same sense of contraction and threat but without the promise of the Renaissance. We talk about a "new normal" or a series, a descending series, of new normals (the Bulletin of the Atomic Scientists calls it "a new abnormal"[12]). Back in the Middle Ages, life seemed normal too. For every reign of terror there was a cathedral. For every plague someone saved Lucretius. We humans, forever adaptable, can convince ourselves that any kind of living is normal.

The term *Middle* Ages implies a beginning and an end. For those who named it, the beginning was the ancient world, which featured the emergence of several brilliant civilizations. After the middle came another brilliant civilization, the Age of Enlightenment, which we flatter ourselves we have inherited. Did any of those enlightened people consider what would come next—i.e., us? The thought was probably an ellipsis (...), the way your voice trails off when you don't know what to say next.

In some religious circles the end was expected at the Millennium. If so, we are now three decades beyond the end, and that could be: we do seem to be living in a world with the light dimmed; a world we no longer understand and whose terrors, real or perceived, have sapped our confidence. We stumble on the path, which is uphill, rocky, hot, and dry. And sometimes wet, for the terrors are contradictory. As are we. We feel a premonition of loss, yet we spend like there's no tomorrow. Our definition of "other" now includes almost everyone beyond our own hearth, yet we pride ourselves on the many virtual friends we have. What is this age?

The Millennialists would say Purgatory or perhaps the Nether Age. Nether Age ("downward") has a nice ring to it, but my dictionary says it's one of the literary words, which I have resolved to eschew. So too is the Crepuscular Age ("shadowy"), both literary and affected. The Twilight Age is cliché and also ambiguous since it usually means evening but can also mean morning. (Had Ronald Reagan declared, "It's twilight in America" instead of "It's morning in America," he would have been more precise but no aircraft carrier would have been named for him.)[13] The Shadowy Age sounds young, and the Age of Shadows asks what cast them, which returns us to fault-finding. The Age of Contradictions could be every age. Since we like air quotes, the "Normal" Age has potential, but what if later generations, missing the irony, dropped the quotes as superfluous?

One name I will not support is The End. I don't think this is the end, and I hope to prove that to you as we go.

We read about the cascading effects of climate change. The Earth is overheating, masses of ice as big as Rhode Island are melting into the seas, and the seas are rising. But what is cascading most today is fear, the one response we cannot afford. It is worse than greed, which is also holding us back. In the future, if this is known as either the Age of Fear or the Age of Greed, we will have failed to live up to our responsibilities.

Perhaps for now I had better stick with ellipsis: the Age of (...).

Misperceived Time

Fear: that is the theme of this book. But I cannot get to fear until I speak about time, for it is our misperception of time that is increasing our fear and preventing us from taking a breath. In Disney's cartoon during the Great Depression, all three pigs were afraid of the wolf, but the provident pig realized he had enough time to build his house of bricks.[14] As a metaphor that is our challenge too. If the provident pig, before his audition, had driven a Humvee and taken all his meals from restaurants and eaten beef there and been a frequent flyer and had his life in his cellphone and been heavily in debt, he would be more like us to start with.

Of course time has two meanings: physical time, kept for us by the NIST Atomic Clock, and perceived time, which has been studied in depth by cognitive psychologists. Perceived time, or I should say misperceived time, is part of our problem in recognizing the threat from climate change and responding to it. We seldom have exactly as much physical time as we think we have; it is usually more or less.

On the one hand is *procrastination,* the feeling, or the wish, that we have more time than we do. As a society we've procrastinated in letting so many years slip by without reducing greenhouse gas emissions—in fact, as

I said in the Introduction, since we've received the first warnings we've doubled emissions.

But now that the horrific effects of an overheated atmosphere have begun to be seen, suddenly we feel *anticipation,* an error in the opposite direction: the paralysis produced by misperceiving the danger as imminent, as here already, disregarding the time still available to prevent even further pollution of the atmosphere and to deal with the worst of its consequences. *We're doomed,* goes this response: *The barbarians are at the gate.*

I wish to be careful of what I say. The change is real, and it will be as bad as reported or worse. The unbalancing of climate systems is happening faster than predicted. Yet the bulk of the harm will develop gradually, however widespread it proves, however marked by natural disaster. Anticipation must not make us late to respond.

Most predictions about climate change end at the year 2100, although scientists warn that because carbon persists in the atmosphere, the climate will be less welcoming for centuries. Still, 2100 is almost eighty years from now, a lifetime, something like thirty thousand days. Eighty years ago was World War II. If we think of what humans have done since then, and we apply those achievements against the exponentially upward trend of progress, there would seem to be nothing we cannot do.

The best of intentions weary with time, however, and unfortunately that too is the threat: climate fatigue. Whatever resolutions we take, we must keep them, consistently, faithfully, stringently, for longer than we know; even when opinion, seizing on any slight improvement in conditions, asserts that the crisis is over. The pendulum of history always swings. Eighty years is four generations, each of which will have to make the same sacrifices, rejecting the same temptations, as we must do; the same or greater.

Everything I Know About Climate Change

Then there is *Time The Handmaid of Perfection.* I
will argue soon that the continual striving to improve,
admirable as it seems in theory, has in practice helped
to despoil our planetary home. As we awaken to the result, it continues to do harm by dragging out our response.

For example, in nearly all the coastal cities of the
world, the seas are rising faster than the measures to
hold them back. That is because those in charge are following processes of decision-making perfected through
years of experience, always with the highest ideals in
mind. A problem is identified, analyzed, quantified, and
understood; solutions are solicited, prepared, submitted,
and sorted; consensus is sought, minority views are con-
sidered, and compromises are made; money is found
(perhaps with conditions); new solutions appear; new
compromises are made; laws are written, consulted, or
litigated...and when, finally, everything is ready and
everyone has been heard from, the watertight gate or
wall must then be built. All politics are local, and responses to the climate problem are nothing if not political.

As a rough rule of thumb, we should figure on an interval of forty years from the first storm surge to the
ribbon-cutting of the barrier. Forty years: time is perfection's *old* handmaid. Disaster recovery moves faster
than that, certainly, but for permanent solutions we are
hamstrung by the ideal process, and it is hard to imagine us, even with a sense of urgency, doing better than
to cut the interval in half.

In contrast to this, and applying our now-familiar
standard of measure, eighty years ago the Pentagon,
still the world's largest office building, was built in sixteen months,[15] and a cargo ship, the SS *Robert E. Peary,*
with 250,000 parts weighing 7,000 tons, went from keel-
laying to launching in 101 hours.[16]

According to some reports, the seas may rise about a foot every decade.[17] The most widely accepted scientific prediction is a range whose lower value is viewed with confidence but whose upper value—the worst case and therefore the thing, to be safe, we need to plan against— is largely unknown.[18] Before the next eighty years passes, people who live in coastal cities will likely have wet feet.

Still, they will be able to live there, and without needing a boat. It is unthinkable that they won't make every effort to keep out the seas. They will have enormous resources to draw upon, for the thing they wish to save is one of the most valuable in the world. Take any typical coastal city—New York, for example. Walk through the streets. Beyond the infinitely precious human lives, keenly attached to their homes, the sheer material creation of the place overwhelms the senses. The financial investment, using the smallest denomination in the TED talk lexicon, must be many trillions of dollars. Skyscrapers, streets, subways, statuary: the city is so big and complex you are amazed by how well it works (drink the tap-water: it's delicious). No one is going to abandon New York without a long, committed struggle.

And there will be time for it. In that time the Wall Street firms, at the southern, wealthier end of Manhattan, could be relocated to higher ground, but the thought is so abhorrent (read "expensive") that present plans call for building a seawall around them. This is known as the "Big U" from its shape,[19] which leaves open the less wealthy top to whatever may come.

Other coastal cities are similarly active. Boston has a plan entitled "Climate Ready Boston" for dealing with a 21-inch rise in the seas.[20] Miami Beach has thought about raising its streets and buildings.[21] So has Venice, Italy, though it has also opted for barriers that rise from the seabed to close its lagoon temporarily during tidal surges—an idea whose realization took fifty-four years,

8

with unexpected side effects.[22] The U.S. Naval Base at Norfolk, Virginia, the world's largest, is replacing its World War II piers with double-deckers.[23]

Of course the cities might fail. In the distant future, if the seas rise enough—if, as some journalists like to say, the new East Coast of the U.S. runs along Interstate 95—the skyscrapers might have to be abandoned. Even with a smaller rise, the many competing stakeholders in a city might be unable to reach a consensus and build the necessary barriers in time to save its subways, waterworks, sewers, and electrical grid, which would be tantamount to losing the whole thing. By the end of the century some cities when viewed from space might be merely scars on the Earth, civilization's strip mines.

What is far more likely, however—I would even say inevitable—is that many coastal towns and settlements will be abandoned, perhaps after destructive storms. These will come often in the future. Each will fuel the debate on whether the taxpayers should continue to indemnify homeowners who persist in living in high-risk areas.

Con will win that debate. The unpopular, gerrymandered National Flood Insurance Program will soon run out of money,[24] and so must FEMA and the Army Corps of Engineers as the disasters keep coming. In the history of the U.S., bailouts are like gowns at the Oscars, worn only once. For that matter I doubt that there is enough concrete in the world to protect the twelve thousand miles of our coastline,[25] even if we could afford it and the terrain and the hydrodynamics were favorable. As the towns suffer catastrophic damage, they will disappear.

Some people may live on the coast without insurance, maintaining their own seawalls; but eventually they too will grow tired of the struggle, for what good is a view if it takes over your life? In the same way those with homes in places prone to wildfire will eventually abandon them. What good is a view (of charred trees at that) if you have to keep your family photo albums always close at hand? And how fair is it to ask emergency responders to risk their lives to save your home when you knew the chance you were taking when you bought it?

By now the fire season is so long, the forests are so hot and dry, and the fires are so destructive that many places previously thought safe are endangered. Fire can come from any direction, across the ground and through the air. As the Earth continues to warm, as the wind rises, large areas in the West will become uninhabitable. Both the buildings themselves and the services that support them will be increasingly hard to justify. Because people are not only proud but sentimental, the uprooting may be a gradual process. But it will happen. Long before the last glacier has melted, long before the rising seas have crested, Nature will have largely reclaimed the mountains.[26]

Besides fire and rain, climate change is bringing other effects. We will live with these every day, knowing they are out there but not thinking much about them until some event reminds us. Conditions will worsen over the century. These include the death of ocean habitat from warming, acidification, and overfishing;[27] widespread drought and desertification, where no food may grow;[28] killer heat waves, dangerous to everyone, not just children and the elderly; urban air pollution from ozone and particulate matter, which is already taking thousands of lives each year;[29] and tropical diseases in once-temperate regions, of which Lyme has been the first invader, with malaria, dengue, and yellow fever certain to follow.

In acknowledgment of the tremendous amount of research on the subject, I should also mention other scenarios that are theoretically possible if unlikely or irremediable. The warmer seas might interfere with photosynthesis by phytoplankton, in which case two-thirds of our oxygen supply would be shut off.[30] If CO_2 concentra-

10

tions rise beyond a certain level, we won't be able to think clearly.[31] If the Gulf Stream stops, Europe will freeze.[32] The 1918 flu pandemic, waiting in the ice, might be released as it melts.[33] The ice-melt itself might cause enormous shifts of weight upon the Earth's surface leading to massive recurring earthquakes.[34] Or this: "We're living in a simulation and it gets shut down"[35] (aka *Life is but a dream*). It's true, if any of these things happen, we're doomed, just like the dinosaurs. But being afraid of them is only going to distract us from doing the things within our control.

What Are We Talking About?

In short, life in the Age of (…) will be harder, sadder, more expensive, and less satisfying in general, unless you are one of its particular victims, in which case it will be catastrophic. But it will be different for each of us, just as it always has been. The future *depends*—a word at the heart of wisdom.

Don't Forget the Other Threats

Two Times Five

So many threats to our future requires a certain amount of triage.[36] I would submit that as far as we can predict, only two have the potential to kill us all, either suddenly or by attrition. These are **Climate Change**, already introduced, and **Nuclear War**. The Bulletin of Atomic Scientists, those keepers of the Doomsday Clock, cite these among the reasons for raising the alarm.[37]

Of these, five would aggravate the risk of the two. These are **Overpopulation**, **Resource Scarcity**, **a General Breakdown in National Sovereignty**, **Automation**, and **Global Economic Collapse**. Since most of the chapters of this book are about climate, in this chapter I will discuss the other existential threat and the five aggravations.

Nuclear War

The Doomsday Clock is now set to 11:58:30—the closest to midnight since 1953, when the hydrogen bomb was tested first by the U.S. and then by the Soviet Union. For context, consider that in 1991, the year the Cold War ended, the clock said 11:43.[38]

Since then, we've assumed that the world was growing safer from the threat of nuclear war. Suddenly that's changed. The Nuclear Non-Proliferation Treaty, in force since 1970 and approved by 190 nations, no longer carries the same weight. North Korea has withdrawn from it. India, Pakistan, and Israel have never joined.[39] In all, nine nations are known to possess nuclear weapons. North Korea possesses thirty warheads and is working to deploy land- and sea-based delivery systems.[40] The U.S. is modernizing its nuclear arsenal.[41] China is both modernizing and growing its arsenal. After the U.S. withdrew from the six-power deal with Iran, it is widely believed that Iran has moved closer to attaining a weapon, something that Israel, an undeclared nuclear power, considers an existential threat. In 2022, amid tensions with the U.S. following Russia's invasion of Ukraine (during which the Russian president often had recourse to nuclear saber-rattling), Russia suspended its participation in New Start, the last strategic arms control treaty in effect between the two countries.[42] The U.S. had already withdrawn from another treaty that bans intermediate-range, land-based nuclear missiles from Europe after gathering intelligence that Russia had been cheating.[43]

Meanwhile, the unthinkable is now becoming thinkable. Nuclear weapons are being built to perform missions once assigned to conventional weapons, as if both nuclear and conventional were tools from the same box.[44] They are part of the ongoing science of how to divert, delay, or fragment a deadly asteroid.[45] At the highest levels of government it was suggested that a nuclear detonation might quench a hurricane.[46]

Missing from these approaches is the recognition that no one can ever use a nuclear weapon, for any purpose, without destroying the anathema against such uses that has existed since Hiroshima and Nagasaki. According to research by anthropologists, one suicide can normalize the practice for an entire community.[47] The same holds true for one nuclear weapon, whose use has always been thought of as suicidal.

The danger has increased too as the unimaginable horror of light, blast, fire, and radiation, the atomization of living matter, the shadows of victims printed on stone,

14

recedes from memory. The terror of the Cold War has passed, and the lessons of that terror have been lost. In contemporary culture, the detonation of the Bomb has taken on the character of a sight gag, entertaining and largely harmless.[48] Against such whimsy it is necessary to recall the actual destructive effects. A 1954 cover of *Life* magazine superimposed a photo of the fireball from a hydrogen bomb test upon the skyline of New York. Because of the scale involved, the Empire State Building and the Statue of Liberty hardly stand out beneath the cosmic dome, within which nothing could live.[49] A single hydrogen bomb. Six nations now have the capability of mounting multiple bombs on intercontinental ballistic missiles.[50]

How soon we forget. How readily we adapt. Here is the means to annihilate the human race overnight. We cannot know all the points of vulnerability, all the possible scenarios of mischance. There have been close calls already, leaked to the public after the fact. Sooner or later it may be that some command and control system will fail, that some leader will miscalculate, that some anomaly of nature or automation will be misinterpreted, that some military cohort will revolt, that some despairing group will seek its own suicide in the destruction of a society. Or worst of all, that some supposedly limited use of these weapons in a regional conflict will involve ever more powerful allies in the kind of vortex of treaty obligations that became World War I. These things are unlikely to happen, but they are becoming more likely. Artificial Intelligence (AI) may make them more likely still. All of us should be clock watchers.

Overpopulation

In 1992 the Union of Concerned Scientists issued a "Warning to Humanity" about the collision course between human beings and the natural world—I referred to this earlier. In their letter, climate change was relegated to a subordinate paragraph among seven existential problems. The main threat was seen as unrestrained population growth.[51] Twenty-five years later, in a second Warning signed by more than fifteen thousand scientists, climate change and overpopulation shared essentially equal concern.[52] During that interval the world population had grown by more than two billion, which should have given the scientists a preview of the consequences.

The population is now about eight billion, and demographers predict that it will grow to about twelve billion by 2100. In theory that would seriously hinder the goal of reducing carbon emissions to "net zero," meaning that the output of carbon sources would be matched or bettered by the absorption of carbon sinks, through photosynthesis and other action. Four billion more people by 2100 will need a lot of food, water, clothing, shelter, energy, and transportation, all of which under present usages will increase carbon emissions. In turn they will be subject to the disasters and hardships of an overheated Earth.

So why the reduced emphasis? We hardly hear about overpopulation anymore.

The obvious answer would be that it hasn't proved to be a problem yet, and there is evidence for that position. According to a United Nations report, during the twenty-five-year interval between the scientists' warnings, while the world was adding two billion people, two billion people emerged from extreme poverty.[53] At first the improvement was mostly in China; but recently it has been seen in every region of the world. If the trend continues—in the same report the U.N. warned that the gains are fragile[54]—we might conclude that population growth, instead of overburdening the planet, has allowed its vast resources to be shared more efficiently and that there will be enough for everyone in the future. Such an outcome would remind us to be skeptical of dire warnings.

16

I'm not convinced by this argument, which seems too easy. Certainly the scientists are right that overpopulation will aggravate climate change: more people means more carbon in the atmosphere as well as greater inertia to shifting from fossil fuels.

And it means more people crowded within a smaller habitable area, which may be the real problem: a critical density of population would have unforeseeable consequences. Think of your house with crowded with permanent guests. So too in communities, regions, nations, and continents. Because we are, as yet, a violent species, the conflicts might turn deadly, on the largest scale.

Some see this as a good thing. Overpopulation, they suggest, might result in a great culling, through warfare or pandemic, famine or thirst; it would be Nature's way of restoring a balance, at the cost of a billion people or so. That sounds like Joseph Stalin, doesn't it: one death is a tragedy, a million deaths is a statistic. Its heartlessness aside, such a view misses the point. If such a calamity occurred, it would be unlikely that the survivors, suitably chastened and now relieved of competitors, would return to their normal lives. Far more likely is that civilization itself would collapse. If the result didn't follow the first of the dire outcomes—a sudden extinction—it would follow the second—a gradual dwindling to nil. One death is a tragedy. A billion deaths may well prove to be the death of all.

Resource Scarcity

The Tragedy of the Commons is a quaint idea that scaled. In 1833, as industrialization was gaining steam and the future seemed boundless, the British economist William Foster Lloyd theorized that the self-interest of individuals could lead them to deplete a resource meant for everyone. His example was free grazing on common land, a local problem.[55] In 1968 Garrett James Hardin, an American ecologist, applied the idea to global over-

17

population, and it has since been applied to depletion of fundamental resources, such as the atmosphere or the oceans.[56] We humans have generally treated the Earth as our own commons, taking what we want without thinking much about what we leave. When tools and methods were primitive, this didn't matter. Now it does. We are depleting resources beyond their ability to recover.

The tragedy extends to the retrograde. We are dumping plastic, a biproduct of oil, into landfills, where it never returns to a natural state, and into the oceans, where it forms a gigantic, persistent, indigestible soup. This wastage may outlive its creators. Many years from now, some other race of intelligent beings may discover, wondrously, the Anthropocene equivalent of a fossil: a piece of plastic imprinted with the logo of an apple.

So what lies ahead? Sooner or later depletion causes shortages. We are seeing that already with water, which has become a critical resource around the world. In the Marshall Islands, for example, where sea-level rise is getting the headlines, the average per capita daily consumption of *fresh* water is only fourteen gallons, for all purposes. (In New York City it is 118.)[57] To find, make, or import it, communities are going to extraordinary lengths. The city of San Diego operates a reverse osmosis plant in which, every day, millions of gallons of seawater are forced through semi-permanent membranes that filter out the salt.[58] The Gulf States of the Middle East operate many such plants, and recent reports suggest that the Emirate of Dubai is working with visionaries to explore the feasibility of towing icebergs into the Persian Gulf from Antarctica,[59] a trip of one hundred or more days requiring massive power, all carbon-burning.

From a distance, the rest of us may think that this is like choosing your poison. To the inhabitants it's about staying in their homes. If oil can be turned into water, they will live with the environmental costs. Again we

18

are reminded that all politics are local, in which the greater good must be seen as equally good for the few.

The possibility of depleting so many raw materials suggests that some strategies of exchange might not work. Renewable energy sources that go idle, like solar and wind, depend on lithium ion batteries to cover the idle periods. Electric cars and almost everything electronic depend on them too. Lithium is rare in the high concentrations that may be extracted cheaply.[60] As ever larger quantities are required—by twelve billion people—the known reserves of high-concentration lithium will be depleted. Suppliers will then need to explore for it or else develop specialized processes for extracting and refining it in low concentrations; either way a lot of machines will be scouring the Earth, not unlike the present hunt for oil. Any way you look at it, the commons will soon be closed.

And when the vital resources run out—or the population thinks they are running out—what then? Either way, prices will rise, people will hoard, and fears will abound, with strife too, perhaps. The 1973 oil embargo caused the price of crude to rise 400 percent and the price of a gallon of gas to rise 43 percent.[61] The psychology was like a run on a bank; the stock market crashed.[62] There were long lines at gas stations with anecdotes of temerity and cravenness typical of panic. The policy of the U.S. and some European countries toward the Middle East veered, a change from which the decades of conflict that have followed became inevitable. All in response to a hint, a foreboding. The reality was something else. Throughout the 70s the embargo caused hardly a tick in rising imports, and domestic production, despite the demand, actually fell.[63] This gives us an idea of what to expect. How will we respond next time? Will we cooperate, will we share? We had better practice, beginning with our attitudes toward others.

Breakdown in National Sovereignty

We are back to the subject of fear again, and the trend is discouraging. In this early part of the Age of (...), when the conditions of life are generally better than at any time in our history, fear and suspicion have gotten under our skin. Some of that is anticipation; some is opportunism—factions seeing their chance, turning lies into dogma. Whatever the cause, people are losing faith in their institutions, and as they do, alliances and federations fall apart.

Late in the last century this was called *Balkanization,* a term derived from the breakup of the former Yugoslavia into smaller states along ethnic and religious lines. Now the phenomenon has swept the world: everyone wants to Leave something. In extreme cases the result is anarchy: when people aren't safe or can't obtain the necessities of life, they will turn on their leaders.

So they have in history; governments fall, and new governments replace them. But any widespread breakdown in national sovereignty now would reduce our chances of survival. Whether we like it or not, the world is divided into governed contiguous nations, which interact with each other to meet their needs and promote the greater good. No lesser or greater authority can supersede theirs.[64] Their number, about two hundred, is right for reaching decisions. Within their borders they have the power to implement those decisions while maintaining an orderly, predictable way of life for their people. Without them other institutions would fail. In their finances and laws our state governments are half-Federal, and municipalities are half-that. Every public utility and private industry receives some advantages from the Federal government, which in turn protects its citizens through regulation (though we dislike regulation until we need it).

Consider the alternative. Without order, the only goal would be immediate survival. Loyalties would be the most Balkanized in thousands of years: self, family, perhaps a loose affiliation with tribe. Conscience would yield to survival and so would hopefulness and so would planning. Desperate groups would break into the nuclear weapons armories. Climate change would be an accepted fact. Life would be reduced to the barest means of subsistence, interrupted by random acts of terror and violence.

If that happened, would anyone care about humans as a species? If someone did, would they be able to do anything about it? If there were no nations, how could there be a census? If people had a rough idea of who lived around them, might they, as the Native Americans did for the Pilgrims, care about their welfare? Or, instead, might they see everyone beyond their campfires as a threat, to be preemptively annihilated. Thanksgiving or Holocaust? I have more questions than answers.

I do believe this: After the collapse of civilization, the best case would be another medieval time, but on a planet stripped of resources. *Then* our species would be prostrate before both Man and Nature. Any rogue occurrence could sweep the last survivors overboard.

Automation

For the moment let's imagine that the Earth isn't overheating, that nuclear weapons are back under strict protocols—or eliminated; and that overpopulation, overconsumption, and national sovereignty aren't problems. We would still have to work out our proper relationship with machines.

We have never known what to make of them. On the one hand, they are marvelous. They are products of the finest qualities of their inventors: intelligence, ingenuity, perseverance, courage, self-sacrifice. They have saved countless millions of lives, made other lives more bearable, and transformed the lives of nearly everyone.[65] Many machines are beautiful in themselves; as

21

tools they help to create beauty. We love and depend on them.

On the other hand, they can be loud, dirty, and noisome—to ride the first trains or work in the first factories was harmful to your health. They require energy, which over the past two centuries has led to despoiling the planet, threatening our very tenure here. When used as weapons they have killed millions—whether on balance more lives have been lost or saved by machines is anyone's guess. We depend on them too much, to the point of addiction.[66] And as we look ahead, advances in their progress toward human capabilities, but with virtually unlimited power and no weaknesses, have prompted fears that machines will rob us of our dominion of the Earth.

The trend seems irreversible. Ultimately, as AI takes off, machines will perform all the work of the world except for the most imaginative, private, or profound.

Even worse, according to some, machines that design machines may doom the human race. Forms of intelligence not yet invented are already conceded to be our masters; as soon as they grow weary of our organic nonsense they will wipe us out. Or, less fancifully, they will meet apparent threats to our (and their) security by attacking them preemptively with nuclear weapons before their human supervisors can stop them.[67]

Those who see such a threat on the horizon are forgetful of the distance to be traveled before we get there, and perhaps they also underestimate our resourcefulness and determination. Machines too must make that journey, which won't be materially accelerated by the repeated doubling of memory and processor speeds, if such a thing too doesn't reach its limit. How will they acquire the totality of our perceptions, the completeness of our consciousness?[68] Our imagination, our sense of beauty, our morality, our devotion? Our reading of character? Even our faults and eccentricities, useful in dealing with the world, are beyond them. What do they know of self-interest? Of nepotism? Of graft?

If the Age of (...) lasts long enough, maybe they will become that capable. But such a possibility is not unlike our asphyxiating when the phytoplankton fail. The practical problem is more immediate than that.

If there were no warming, our civilization would still need to adapt to enormous changes. The two central—I will say existential—questions facing us will be these:

1. When machines do so much of the work, how will we humans make a living?

2. Even if our necessities are assured, how will we find purpose in our lives?

There is room for new thinking on both questions. We might, for example, look at automation as an opportunity rather than a threat. If the slate were blank, how would we *wish* to make a living? What gives us purpose today? Is it only work? Could it not be self-improvement, learning, creation, recreation, joy, love, pleasure? The answers will need be examined critically and realistically. Not everyone wants to write poetry (or essays about the future).

Economic Collapse

As successful as the world's economies have been, the pressure on them is immense. Decades of economic growth are threatened by the tragedy of the Earth-commons and by the loss of jobs to automation. As I will argue in the next chapter, our response to climate change will require reducing economic activity, perhaps for a long time.

Recession, then, as we know it in our capitalist economy, will become the status quo; or perhaps something more severe. About two-thirds of the world's population are of working age.[69] If machines perform half the labor of humanity in 2100—which seems a low estimate—four billion people would be out of work, unable to make a living. Credit would be tight, but inflation would be high—scarcity's public face—with resultingly high interest rates. We had a mild taste of this during the 1970s, when millions of Americans lost their jobs in the (ultimately successful) effort to reduce inflation. What I have in mind will be much worse: higher unemployment, higher inflation, lower productivity.

Something can be done about this, up to a point, as long as institutions exist to do it. Those, nostalgic for simpler days, who wish to abolish the income tax, disestablish the Federal Reserve System, and break up Wall Street should think of the consequences. Without credit bureaus there would be no credit cards. Without the Federal Deposit Insurance Corporation, no bank accounts would be insured. If no institution set the cost of money, usury rates would soar. We cannot have cellphones without a network behind them, nor can we live in any semblance of modern life without the enormous financial infrastructure that makes an economy work.

The hard times ahead will be immeasurably harder if we cannot function as a society.

Let's Be Realistic

This chapter ends with a larger point, about practicality in general. All credible scientists warn that the threat from global warming is urgent. In April 2022 the Intergovernmental Panel on Climate Change (IPCC), the U.N. body assigned to evaluate the best science on the subject, reported that to hold warming to 1.5 degrees Celsius, one of the goals of the Paris Agreement, carbon emissions would need to be reduced sharply by 2030.[70] As reported in the media, anything later would be too late.[71] That may be. But the changes we need are not going to happen as soon as the scientists advise; nor to begin to happen. It's not possible physically, psychologically, politically, or institutionally. The projections of the harm that may be done as well as the remedies that may be applied should take this into account.

When the danger is urgent, slowing down seems counterintuitive, but in this case it's essential. Before we deal with a problem, we need to understand it. The proximate cause of climate change is well-known: too much carbon! But there's a deeper cause. Understanding, accepting, and addressing that will take us far beyond 2030.

The Great Binge

Always Upward

What the future asks of us it has asked before—not in the details but in the terms. During the Middle Ages, despite invasion, plague, and famine, generations of craftsmen without our advanced knowledge built the magnificent cathedrals of Europe. The work, unfinished during many a lifetime, was carried on by religious faith and filial obligation.

Something else moved them. It is the nature of living things to improve their condition, overcoming obstacles with ingenuity and hard work. All of us humans, possessing imagination as well, are born to aspire. Despite a few pauses, our history has been one of progress. Hunter-gatherers turned to planting and herding; users of stone tools discovered bronze. About the time of the American Revolution, noted for its breakthrough in political philosophy, material inventions took off. The steam engine in its many applications offered astonishingly higher levels of productivity and quality. Iron and steel; interchangeable parts; distributed electricity; the telephone, radio, and TV; the car and the airplane...always upward.

The heart of invention is noble. We've seen that in our own day. Someone as yet unknown to the world has an idea, perhaps beyond his usual experience, and he gives himself up to it, constantly trying and failing and trying again. He takes risks. He goes into debt. He transcends himself. Restless to improve, always he strives for perfection, often redefining what perfection means. Whatever fame and wealth he may earn, we think he has done civilization good service.

One Cause

And that's the problem: he *has* done civilization good service. When the first shovelful of coal went into the first firebox, we humans were set on the path that must lead to a hotter planet. For more than two centuries, mostly without thinking about it, we've been on the Great Binge. As the eminent scientists put it in their second warning, we've improved our condition by creating and wedding ourselves to "an economy rooted in growth."[72]

Growth means consumption—it didn't have to be so, but it is[73]—and now we will do anything to protect our sources of supply. In that resolve it has seemed that the Earth would always provide for us and that if a needed material ran out, another would take its place or we would devise a way around it. Each period when we worry about dwindling resources is followed by another of unbridled use. So we continue to take and consume, discarding the leftovers, which might have been shared. In fact our way of life, humanity's growth-based, market capitalism fostered by governments of all kinds, has become too perfect,[74] a victim of our aspirations.

Step back from the debate about how to respond to climate change. Consider the fundamental logic:

☞ We have despoiled the Earth of natural resources.

☞ Our way of life is based on always seeking more.

Given enough time and progress, how can the first condition not have resulted from the second?

More. More of everything. Money earns interest. Stocks pay dividends. Investors demand growth. Markets react merely to the rumors of growth or its oppo-

28

site. Your local bakery may sell the same number of croissants every year, but big business by its nature seeks to become bigger business, and every entrepreneur dreams of turning the idea in his garage into a fabulous IPO. Did fossil fuels fuel the Industrial Age? Did the Industrial Age introduce the demand for windfall profits? It's all tied up together.

The most obvious proof of this walks past our windows. Even as late as the 1970s the average citizen of a Western nation was at an ideal body weight; all the old movies show how lean we were. Since then we have been practiced upon by an enormously powerful, sophisticated, and well-developed sales and marketing machine whose purpose is to increase not only the number of its consumers but also the caloric consumption of each—without regard to health. Besides ubiquitous advertising, food and drink companies have achieved their goal through two hooks: they have made processed food less expensive and more convenient than cooking from scratch, and they have made it irresistibly sweet or salty or both ("Betcha can't eat just one," the potato chip manufacturer taunted us[75]). While this was going on, people began dining out more often, subject to the same marketing ambitions.

Forty years of overeating: of course we're fat. The average body weight has shifted so far toward the obese—and the obesity rates are staggering[76]—that, perhaps in defense of our self-regard, the ideal has shifted with it. Beauty is now an endomorph, big and strong with well distributed fat. To decry the change, even on medical grounds, is to invite accusations of body shaming. But the medical grounds are compelling: among the items of a long list, obesity causes depression, pain, heart attack, stroke, gallbladder disease, some cancers, and Type 2 diabetes—in other words, a low-quality life, terribly difficult to treat and probably ending too soon.[77]

It's not only food. As court cases have revealed, opioid addiction mushroomed after aggressive marketing by some pharmaceutical companies. The vaping of THC, the active ingredient of marijuana, is linked to death from lung disorders; entrepreneurs—that is, people with aspirations—promoted e-cigarettes and pressed for relaxation in the drug laws before the research had been completed.[78]

Nor is it only our bodies. The Great Binge has been about overconsumption in general. We brew hot coffee that we then chill with ice; we heat and air-condition buildings at the same time. Cars have gotten bigger, heavier, and higher-tech, and there are far more of them than one in every garage (ditto chickens in every pot: Herbert Hoover's 1928 campaign pledge has been more than fulfilled). Servers have become server farms; hard-drives have become The Cloud. The college dorm room of faithful cinderblock has grown into a luxury apartment, managed by real estate professionals for students treated as customers; no self-respecting university fitness center is without its climbing wall and lazy river. The in-home theater is becoming a staple of the middle class. Monday night football, Thursday night football, football in London and Mexico City. October baseball in November; ice hockey in June. Bigger is better. *Enough* fell out of fashion ages hence, replaced by *more.* It's what we've demanded.

Along with material goods we've perfected our institutions. Consider the global corporation on its "campus": not only are its products marvelous and constantly improving, the entity itself is a model of administration. It operates effortlessly in foreign countries; it knows all about ships, airplanes, trucks, and pipelines. It imposes its will wherever it goes—Clausewitz's definition of success in war. Profitable each year and ambitious to be more so, it is not without heart. It puts its name not only on football stadiums but also on laboratories. When disasters happen, especially those caused by its operations, it partners with governments to mitigate the damage and relieve the suffering, up to the point of profitability.

They've all been too successful, all the institutions—industry, non-profits, universities, advocacy groups. The corporate world is not the cause of climate change but rather its most excellent manifestation. As we binged we admired those who provided the means. It is only now, when the results are seen, that we regret. How ironic if our success were to remove us from the Earth. Whether we have the will—we and our children and their children—to unbuild the cathedrals is the vital question.

One Cure

How to fix it? Eventually *more* will have to be replaced by *enough,* which indigenous tribes understood as the true bounty of the Earth. In the foreseeable future, we will have to do *less.* Not more with less—that confidence game—but less with less. Growth will have to give way to scaling back.

In many cases that means preferring the old to the new. It has been our desire for the new, after all, that has spurred economic growth. We hardly notice old objects, but if they were affordable, well-built, still useful, and pleasant to the eye, maybe we would look again. As Cuba has shown, a society can get around quite well in used cars. *Less* wouldn't feel like such a pinch if it were supplemented by what has been reused, recycled, or repurposed.

Let's return to the inexorable logic:

☞ We have despoiled the Earth of natural resources.

☞ Our way of life is based on always seeking more.

This leads to one conclusion:

☞ To save our home, we must change our way of life.

The solution is as simple to state and hard to do as that. Hard, because of the economic consequences: the job losses from underconsumption and automation will be additive, causing widespread, long-lasting unemployment. Think of a Great Depression for which there cannot be a New Deal, on top of a less bountiful planet subject to frequent natural disaster. No matter how long we think we can put it off, a reckoning will come, and misery will follow. The point is there's no other way. Changing our way of life means changing not just how business is done, what money means, and what resources symbolize. It means changing our laws, possibly our systems of government and economics. It means changing how we relate to each other and what we demand of our leaders. To some extent, still to be worked out, it means lowering our expectations.

We can make these changes deliberately, or they will be imposed on us, to our cost.

Opposition

No One Likes to Downsize

Humor me again, please: imagine that you are a candidate for president of the United States. At a key debate you are asked what your campaign stands for. "Doing less," you reply. "I'm going to cut back. We're going to live on what we have."

"You mean doing more with less," the moderator helpfully prompts.

"Nope. Doing less with less."

How do you think your candidacy would do?

Actually, history provides an answer. Jimmy Carter was elected president in a grassroots reaction to the abuses of power by his predecessors. For three months he enjoyed a honeymoon in office, with an approval rating in the eighties.[79] Then, overconfident, he asked the American people for sacrifice. Addressing them on national TV in a speech he termed "an unpleasant talk with you," he called for meeting the energy crisis by ending our dependence on foreign oil, a cause he termed "the moral equivalent of war." The alternative, he said, "may be a national catastrophe."[80] Among the ten principles of his action plan, conservation—using *less*—was the main idea. It landed with a thud. Although Americans had been inconvenienced by the oil embargo of 1973, the pain had passed, and they didn't want tough love from their president. Carter's popularity fell from that moment. The "moral equivalent of war" was derided in the media by its acronym, MEOW.[81] Other crises in his presidency, particularly the Iran hostage standoff and the failed desert rescue attempt, reinforced the impression of a man too good to be powerful (exactly the qualities voters had said they wanted when they elected him), but the impression was formed by the energy speech, and Carter's defeat by Ronald Reagan, a reactionary candidate, after the one term in office came as no surprise.[82]

No one likes to downsize. Smaller is un-American. Ambition requires growth: the U.S. wouldn't have fifty states and fourteen territories today without it. Corporations grow or else are abandoned by investors, customers, even employees. Cities incorporate suburbs. Universities operate hospitals, which operate gift shops, which operate coffee stands, where every barista knows whether the business is growing or failing. That is the choice: grow or fail.

Rather, that *was* the choice. Now the choice is to live within our share of the renewable resources of the Earth or contribute to the likely end of humans and a million other species.

However true that is—and I hope this essay convinces you that the logic is irrefutable—countless people and institutions will oppose it. Partly out of self-interest and partly because it contradicts what they have always believed about the future; that is, its inexhaustible abundance.

Self-Interest

But mostly because of self-interest. I could put that more kindly: the future will require sacrifice from everyone, and while such a thing is easy to say in the abstract, the pain will be specific, widespread, and real:

Worker: *Your recession will cost me my job.*

Suburban homeowner: *And my home.*

Farmer: *Without price supports I'll lose my farm.*

Parent: *I need all our cars to raise my kids.*

Car Dealer: *We can solve the climate problem with electric cars.*

Cattleman: *Stop selling beeves? We've ranched this land since the Indian Wars.*

Food Wholesaler: *Our customers want choice; eating local is too limiting.*

The Military: *Without the most competitive weapons we cannot defend our country.*

The list might go on and on.

People who oppose the idea of *less* can go a long way toward preventing it. They can vote, they can spend their money to sway elections, they can be elected themselves, they can gerrymander districts, they can pack the courts, they can encourage the police to protect property above life. Apartheid in South Africa showed that a small minority can frustrate the will of the majority by ruthless coercion—for decades. Abolish fossil fuel drilling and mining? You might as easily abolish football.

Institutions are thought to act more rationally than individuals; their decisions, they proudly say, are driven by data. Compared to the data they hold dear, each company's share of the stewardship of the Earth is hard to measure, relatively small, and invisible to the bottom line. When profits mean so much, it takes a brave CEO to propose leaving resources in the ground. And no CEO, even with his firm's carbon footprint before him, wants to cut back production. In truth even the CEOs of oil and gas companies don't really know what to do now. The safe thing is to keep drilling. Like many of our political leaders, they hope the problem will be solved by innovation, by smart growth. Let the market work this out.

Yet the market too opposes *less*. It recognizes return on investment—mostly present return, for it is hardly more foresighted than its members. As long as a company reports greater profits with each business cycle, the market will reward it. In the striving for perfection, the goal has no limit, though the resources run out or the costs run up.

Thirty-five years after the first public warnings, forty-six years after Carter's speech (whose adoption by the country would have gone a long way toward solving the same problem), the opposition to effective action is everywhere.

The Immovable Object

Perhaps the greatest resistance will come from inertia; from the magnitude and character of the sacrifices required. Humans have been on this path in general since the invention of the wheel but most particularly since the steam engine. Billions of people depend on big and expensive carbon-emitting machines. Our way of life simply cannot be changed on the timescale that we usually associate with managing crises. John Kerry has called it "trying to turn around the largest oil tanker ever built."[83] That's an apt image, but even so it doesn't begin to address the size of the problem; perhaps the problem is too great for metaphor.

Think about what's involved—pick any emission source. In the U.S. alone it will be impossible to convert 300 million vehicles to electric drive or something better overnight. The cattle industry will not disappear in a new kind of Prohibition. Travelers won't soon abandon airplanes for trains and buses. Around the world the obstacles are proportionally greater. Palm oil is so widely used in food processing that palm tree plantations will continue to take over the rainforests of Indonesia. Brazil won't stop logging and burning in the Amazon as long as people can profit from it. It will be decades be-

Fear

fore the thousands of fossil fuel power plants will be de-
commissioned.[84] (If there is fossil fuel in the ground, it's
a safe bet that someone will burn it—the Alberta tar
sands will continue to be mined long after the waves lap
our coastal cities.)

We like to use one of the hopeful slogans of conservation: "Every little bit helps." True, but right now every
little bit *that hurts* needs to be replaced, and the scale
itself of such an effort is a challenge: there are too many
little bits. Think how difficult it has been to distribute
lifesaving medications and vaccines to remote areas of
the world and to overcome taboos against them. Who,
then, will convince impoverished Africans to turn in
their charcoal stoves? Resistance to change is universal,
and as nations develop their industries even during the
climate crisis, that development will come with built-in
inertia: as we are taking cars off the road in the United
States, they will be adding them in India.

Fear

In a greater sense we are still making sense of the
threat. When I was in the Navy this was called *recognition differential*—the amount of time required for the
sonar operator to recognize, for example, that the echo
on his scope might be an enemy submarine, taking into
account the contributions of fatigue, boredom, horse-
play, training, and gofer duties as well as the ordinary
difficulty of detecting a relatively small metal object in a
great and noisy sea. Had we been in a shooting war, it
would have included fear.

In fact all of us are subject to this behavior and more
often than we think. If we knock a dish to the floor. If
we step on something sharp. If we see, entering an intersection, a car on the crossroad running the red light.
There is always a moment of shock, a *this can't be happening,* before we wake up to the altered situation and
begin to form our response. (Since mine is meant to be a

hopeful book, I should add: a marriage proposal, a pro-
motion at work, and finding a gift from last year when
we unpack our Christmas stocking.)

The world at large is suffering from recognition differential over the climate problem. Because the alert
was first given in 1988, that's taking a long time to see
the submarine. But remember that we're not talking
about detecting the threat, which has already been done
by the scientists, but the time required, the delay if you
will, while we recognize and accept it. In this case ac-
ceptance may take much longer than recognition.

So even now, when the scientific evidence is over-
whelming and so much damage is already apparent,
some people deny the threat, and many millions of others try to ignore it. In response to this, the environmental community, in despair at our failure to heed the
warnings and as the window closes for taking effective
measures to prevent the worst, raise their voices. They
make no bones about trying to scare us. They conflate
the present real danger with the future possibility of
human extinction. Using the goal of 1.5 degrees Celsius,
they describe as a monolithic, looming catastrophe—a
yes or *no, inhabitable* or *uninhabitable*—an enormously
complex, varied, irregular change in our environment
over the vast area that constitutes the Earth and which
won't be fully manifested for decades. In effect they reduce climate change to an allegory we're living in. If I
were they I'd probably do the same. Yet the real problem is that we haven't all recognized the sonar echo yet.

Whether the environmentalists' approach is effective—whether fear will move us to earlier action, we are
certainly afraid. We have responded to the warnings
with fear's biproducts: denial, resignation, excuse-
making, and finger-pointing. The process of getting
through recognition differential is much more complicated than, say, the five stages of Kubler-Ross, but no
doubt there are similarities. Even if we know there's a
submarine ahead of us, we are reluctant to zig-zag. It's a

short step from there to rejecting the idea of the submarine altogether.

Blame

Blaming others is a kind of denial, a way of preemptively denying our own responsibility. So we blame specific individuals, industries, nations, and generations. The generational blame is natural but absurd: everyone above a certain youthful age is wearing mink.[85] Lost in that accusation is the underlying cause, which affects all of us. There is no point in blaming our dead forebears; most of them had no idea what their way of life was doing, any more than smokers in the fifties knew that they were shortening their lives. We who are alive are responsible but not guilty. All of us live as best we can. Even young people cherish their cars and board their airplanes.

Climate Fatigue

Instead of saying "We will never forgive you," we should be saying "What can we do together?" and "What can I myself do?" The rest of this book—this hopeful book—will try to answer those and other important questions.

Opening minds won't be easy. Changing habits will be hard. Accepting *less* may be insuperable. Doing these things consistently and rationally for as long as required, though the siren song of *more* will always be out there and may for brief periods of time seem both plausible and harmless, exhausts the imagination.

Tireless devotion isn't our legacy. It remains to be seen whether it is part of our character. After Pearl Harbor Americans grudgingly accepted government rationing—"There's a war on," they said to each other. People could still eat meat, eggs, and cheese, only less of them. They could still drive their cars if they had a good reason (there were no new cars, however, and no rubber for tires). Train and bus travel were limited only by capacity. Flying was mostly unavailable, but few Americans flew then anyway. All in all, and particularly compared to the horrors in Europe, citizens on the home front weren't terribly deprived. Yet as the Allies closed in on victory, the American public wearied of *less,* and the rationing of gasoline was actually ended even before the Japanese surrendered. The pent-up demand of only four years led to unprecedented growth and a taste for luxury afterwards.

In the same way, in 1974, after the Arab oil boycott and the long gas lines in America, Congress passed a law setting the national speed limit to 55 m.p.h. The goal was the same as Jimmy Carter's, three years later: to wean us off foreign oil. From the first the limit was widely violated, although we kept it in place for twenty-one years until, flush with the global oil surplus, we repealed it.[86] Since then, we've put more than half the total amount of carbon dioxide into the atmosphere.

So *patience.* Assuming that we adopt them as we must, the constraints on our way of life, on growth, will be with us for a long time. We will grow weary of them, but no matter how tempted we may be, we mustn't return to our wasteful patterns of consumption. Though much of the damage has been done already, it could be worse, and it *will be* worse if we waver. Dealing with climate change must be a united, non-partisan goal. Among the other proverbs we must teach our children, to teach their children, is *waste not, want not.* Conservation must cease being so remarkable that it gets a name. We falter at our peril.

Eight Billion Approaches
To a Solution

How to Proceed: Six First Steps

Books like this can be full of grand ideas but short on policy. *Where's the beef?* you might ask, as Wendy's and Walter Mondale did (a question that must become literally true if we are to curb emissions of methane).[87] Nor is it enough to criticize the ideas of others without offering alternatives. Here, then, is my program for meeting the future: its threats to our survival, its imperative to change our way of life, specifically its imperative to embrace *less.*

The point is to start.

Stop Digging the Hole Deeper

Several industries, notably energy and cattle, have succeeded in obtaining public preferences for a variety of products whose use in any amount will further damage the environment. As a first step, those preferences should be ended.

Ending the oil and gas depletion allowance seems obvious if we are to disincentivize the burning of carbon. But since transportation is the largest cause of carbon emissions,[88] we should look at all possible remedies. The U.S. Highway Trust Fund, which pays for the roads, has operated at a deficit for years; this is because the federal excise tax on gasoline (18.4 cents/gallon) and diesel (24.4 cents/gallon) hasn't been raised since 1993 nor is it indexed to inflation.[89] Besides the problem of fairness— that non-drivers through their income taxes are subsidizing the lifestyles of drivers—keeping the tax artificially low hides the true cost of vehicular transportation and therefore encourages overuse. If the tax were raised to cover the full costs, fuel would be more expensive, and people would think twice about driving so much and in such big cars and trucks.

Even better would be to tax carbon at every step of its journey from the ground to the atmosphere. As part of that effort, individuals and institutions should have an easy tool to track their carbon footprints, a monthly bill with a standardized accounting, *in metric,*[90] that could be aggregated at the state and Federal level and shared with other countries.

Subsidies to the cattle industry hurt the environment too and by undercutting actual costs—a familiar pattern—leads us to eat too much red meat, which hurts our health. Price supports and crop insurance have tied farmers to government with a Gordian knot; however well or badly these policies may have worked in the past, in the Age of (...) farmers must be self-reliant.[91] The market is hardly omniscient, but the law of supply and demand is still the best agent for changing behavior. All products should compete fairly. Even renewable energy shouldn't be given preference, since better sources might be just around the corner.

Finally, we need to set a realistic policy about emergency relief and indemnification. People who build homes in untenable places should be prepared to take their chances and bear the costs.

Start the Descent

Letting the economy fly on until automation or the climate brings it down would be a mistake. We need to talk openly about the future and the changes required to live within our means, which is no different than living in harmony with the Earth. Even more urgently, we need to balance the budget at all levels, governmental and household. John Maynard Keynes argued—and history has shown—that in a market economy, nations should run deficits to stimulate growth and accumulate surpluses after they achieve it. The economy I envision would stop setting growth as a goal, except that which occurs inevitably with added population.

Why do these things?

First, we will need the money later. That is, our children and their children will need it. The costs of saving cities and securing food and water supplies and relocating displaced people will be unimaginably high, at a time when incomes can only go down. As of this writing the U.S. national debt is more than 31.5 trillion dollars, *roughly eighty thousand dollars per citizen,* and approaching one hundred percent of the Gross Domestic Product (GDP), the usual standard of measurement;[92] private-sector debt, held by households, non-financial corporations, and non-profits that serve households, is roughly twice that of GDP.[93] All that debt is carried in the expectation of future revenue: when revenue must fall, running large deficits is no different than running a Ponzi scheme.

Second, we need to overcome inertia. Everyone decries the evils of indebtedness, but households and governments keep spending. Just as with the idea of confronting climate change, the idea of tightening budgets makes us run after the jingle of the ice cream truck instead. Talking about the coming recession and then committing to it through sustained budget cuts would bring on the reality before the reality becomes catastrophic. After all, the economy is still growing. While we are prosperous is the time to prepare for adversity.

Because markets are like a murmuration of starlings (with a few contrarians), the first wingover will set the new direction. Let the recession begin.

Improve the Physical Infrastructure

Yes, repair and improve the material foundations of a failing way of life, some of which will obsolesce or be under water in the future. If this seems improvident, recall the importance of continuity. Above all else, we must hold our civilization together. The transition to *less* must be as graceful as we can make it, or we risk losing the future to disorder. In that effort the Broken Windows Theory, whatever its flaws in policing, is valid as sociology.

An important part of our self-confidence as Americans is the idea that our values are superior to those of other nations: our tap-water is good to drink, and our buildings don't collapse in earthquakes.[94] On the other hand, when a bridge falls or a train derails, we feel a sense of national shame, that we ourselves have failed. For a number of reasons our self-confidence has taken a hit recently. We've been here before. During the Great Depression of the 1930s, it seemed to many that the basis for their faith in the country had crumbled. In the future we too will feel the loss of faith, and more of us will feel it. If we are to meet the threats fearlessly and with resolve, we must regain our storied can-do attitude.

The element of morale is important, but there's also a practical reason to take this step. As with most material possessions, it will be cheaper and less disruptive to maintain the country's infrastructure as we go rather than to deal with it when it's past repairing. In October 2021 the U.S. passed a bill to meet this goal, one of the largest in our history—1.2 trillion dollars. It was a major bipartisan achievement that will fund maintenance of the present infrastructure and also invest in technologies—the internet, power grids—that will help define the infrastructure of the future.[95]

I would argue that more needs to be done.

46

Improve the Physical Infrastructure

Infrastructure includes many things that don't move people, but in carrying out a national program we will determine the future of transportation. What should that look like? The transition, while it emphasizes new possibilities, must respect its legacy. We will need to spend money on bridges and highways, which will be with us for many years (some highways will need to be raised to keep them dry). And on railroads, old and new; it's a scandal that the most innovative nation in the world should have such a dilapidated rail system. And on seaports and ships, which move large goods economically. And on airports and airplanes, for we have grown accustomed to flying, despite their outsized carbon footprints.[96]

But each of these modes of transportation, the sector that pollutes the atmosphere more than any other human activity, must compete on the basis of its usefulness, adaptability, and cost, which includes the cost of carbon. Time is running out for them to show that they can operate without further damaging the planet. We know what damage cars and trucks do and what the automakers have in mind for them, consistent with profitability. The shipping industry, which contributes three percent of global carbon emissions,[97] is struggling with the difficulties of curbing emissions from its propulsion plants. So is the aviation industry, which moves smaller loads at a disproportionately higher environmental cost (I'll have more to say about airplanes when I urge you to fly less).

I am betting on trains. Not only are they efficient and relatively green, they can be made even more so. Electric-drive trains will realize the benefit of renewable energy sources on a large scale, as if several hundred gas-guzzlers were instantly converted. Trains that run on magnetic levitation (Maglev)[98] or in partly evacuated tubes or tunnels (Vactrain)[99] will build on that savings to deliver passengers or cargo at higher speeds while reducing power consumption overall.

Higher speeds. In general, however, embracing *less* in our future infrastructure means that life will slow down; we need to get used to the idea. Just-in-time and same-day deliveries are unsustainable. So too are the big box stores, to which people drive long distances and park on immense asphalt lots where food could be growing instead. Reductions in demand will get these off the landscape and many of the giant trucks off the road. But trains will flourish. Except in emergencies or for national defense, a day's travel ought to be about as far as Amtrak and its competitors—I hope a lot of competitors—can take us.

It's likely for financial reasons that families won't live so far apart. In fact, as the world adapts to automation, many societies may return to the practice of housing multiple generations together, pooling their resources, including childcare. Living within a day's convenient train ride is almost as good. Whatever may be lost in opportunity will be paid back in more meaningful relationships. We keep saying today that family is everything, yet our culture scatters families, and cyberspace unites them only in an unsatisfying semblance of real contact. Wouldn't we be happier otherwise?

Reclaim Arable Land

It's not clear—there are too many variables—whether the land left to us for agriculture from the shifting patterns of heat, drought, rain, and rising seas will support the Earth's larger population. Even in the United States, it will depend. Therefore, at least as a hedge, we must end suburban sprawl.

As many studies have shown, this must be done early on, before utilities are laid, before farmland is rezoned, before it is even broken into parcels. All these things are subject to policy, which now must change. It's *not* the economy, stupid. Except on the hustings, elected officials are just as baffled by economic forces as the rest of

us, and many a heralded announcement of new jobs
proves to be so expensive in tax breaks and other incen-
tives that the employees concerned might as well be on
the public payroll. It's the Earth, my friend, and our
leaders are writing a record for history to read. If we
will reward those who act from that knowledge, in the
future we may see developments bulldozed and food
grown where swimming pools used to be.

Perhaps more than any other remedy, this will take
time. But it may happen by itself. If home prices don't
recover from the coming recession, the housing market
won't be able to unload the surplus. Eventually many
developments will provide a better return once again as
farms. However long it takes and however irregular the
transition, when the world needs arable land, ghost vil-
lages will be reclaimed and readied for the till.

Revitalize Cities

It may well be that the green revolution will begin in
the cities and spread outward. Cities have a continuing
interest in reducing their carbon footprints and in
adapting to the local effects of climate change. The authority of each City Hall is credible, with enough resources to fund its mandates. Being accountable at close
range, it can be both sensitive and pragmatic, respon-
sive and protective.

Among the resources it may draw upon are hundreds of acres of rooftops, where solar panels might be
installed, and millions of niches in superstructure,
where trees and other plants might capture carbon.[100]
To augment local sources of food, fruits and vegetables
might be grown there too, an urban reprisal of Victory
Gardens.

All this assumes a change not in purpose but in emphasis. As part of that, cities will want to reduce vehicular traffic. Private cars and ridesharing should give way
to public transportation, biking, and walking. This is

not only possible but easily done, within limits, over time. Cities outside the U.S., like Amsterdam and Copenhagen, have shown the way. Reducing the number of trucks will be harder. It may be that specialized delivery systems will replace them (perhaps a network of nearly silent Maglev trains on raised rails) or that, as the economy contracts and the big box stores fail, trucks will become smaller. A city can be efficient and also human, if humans are given priority.

The revitalization of any city must depend upon, as it must foster, social justice. A vibrant city where millions of people lead lives of their choosing seems a long way from conditions today. Walk through the worst neighborhoods of any city at dawn, past the bundled homeless figures, and you will feel how far cities still have to go. Reversing urban flight will help...as long as the poor aren't squeezed out. When all the schools are good; when all the streets are clean and safe and well shaded by trees; when all forms of discrimination are cast off; and with all its amenities and attractions, who wouldn't want to live in a city?

Practice Forehandedness[101]

About the last thing people do in a crisis is look ahead. When the present is bad enough, no one thinks to prepare for worse.

Then too Americans aren't forehanded by nature. When we solve problems, it is usually at the last minute, too late to prevent the consequences. In our culture we value improvisation over planning, the daring escape over avoiding the danger in the first place. Generally, we are unprepared for even the greatest and most obvious threats. It's rare that our leaders will even talk about them.

We need to talk now. To begin with, we need to devise ways of talking beyond the staged rally, the stilted town-hall meeting, the polemics of cable news and call-in radio, and the propagandizing of social media. We need to talk in person, in depth, with humility. Think how valuable our conversation would be before the next trillion-dollar infrastructure bill.

How Not to Proceed: Three Fallacies

The Sanctuary

From time immemorial humans have taken cover. One hundred thousand years ago there were cave-dwellers. More recently we built fallout shelters, and when the Y2K bug seemed Apocalyptic, many of us re-commissioned them, only to eat the canned beans and remove the guns when the bug was fixed in time.[102] Now that climate change has raised a new fear, some of us are reading the scientific predictions with maps in hand: seeking the regions safest from the consequences; planning sanctuaries, now or in the future.

None of these efforts, nor their motives, are to be despised. People want to be safe, and they want their descendants to be safe. A thousand years from now, if only a few humans survive and only then because their forbears took cover, anyone reading this book will laugh at my naïveté.

Nevertheless, I would respectfully offer a dissent. I believe that the reasoning behind the sanctuary, like so much of the doomsday literature that exalts it, is full of holes.

First, in the great majority of cases, the sanctuary could not be built. The desired location, often in a foreign wilderness, would be problematic because the roads wouldn't exist to bring in the raw materials of construction, nor would the current inhabitants allow it if they did. An island supplied by ship might work, but what about the rising seas and the vulnerability to powerful storms? You'd need the Seabees.

If somehow the raw materials could be supplied, during construction you would want to conceal the purpose and location of the project. But that wouldn't be possible. Either you would have to bring in your own construction and security crews, causing further offense, or you would have to rely on a local workforce. In either case the secret would get out, and suddenly you would find yourself with a lot of new friends. In the 1950s and 60s people knew when a neighbor had built a fallout shelter in their basement. Imagine the difficulty here when the construction site would be acres of open ground.

If somehow the sanctuary could be built in secret—or built and forgotten, after the early interest—it couldn't be sustained, largely because of the same limitations of the wilderness. The problem wouldn't be the bare necessities of life but the finished goods and services of civilization. It would be unrealistic to think that machines could be serviced or communications maintained or raw materials finished or adequate medical treatment provided or you and your family prevented from going stir-crazy. Unless it grew into a society itself, satisfying its own needs, the sanctuary must fall to ruins. If you want civilization, you have it already: in our cities.

But the greatest threat to the sanctuary would be its vulnerability. From the first groundbreaking you would feel uneasy about this. If law and order collapsed, untold numbers of people would demand that you admit them—imagine our present border crisis with vastly more refugees aiming at a much smaller objective. A security force could not keep them out, for two reasons: first because, I hope, you would not want them harmed; and next and in any event, because the refugees would greatly outnumber the defenders. If the demand for admittance turned deadly, as it might, only an army with all its firepower could prevail—but in that case the national sovereignty must still exist, because only nations have armies, so why not stay home and trust in your own government to protect you?

In any case the timing is wrong. To buy the land and build the sanctuary, you would want to act now, before the rush—and before the deluge. But the collapse of societies, if there is one, will fall not on you nor probably on your children but on their children and their children's children (eighty years, four generations). *You* may see the sanctuary as wise and providential, a house of bricks. But your children will have their own opinions, probably different than yours, and even if they agreed with you it's not likely they would maintain through their lifetimes the expense and effort of the sanctuary when there was no immediate necessity. They'll want to live somewhere else, with less constraint, even as the seas rise. The sanctuary will exist in family lore as a white elephant...until, suddenly, it is needed, when they might find that it had been taken over by strangers.

Finally, be reasonable: you wouldn't want to live there, and neither would your descendants. They might get used to a subsistence life in the wilderness, but not if they must defend it and never leave. Instead of a sanctuary, you would be better advised to spend your efforts saving what you have. The easiest, most effective way to do that is to live in harmony with the Earth and to uphold the authority of an honest government.

Space Locusts

Then there are the fantasists who would decamp to a space ship or another planet. Many of the objections to the Earthly sanctuary would apply to one in the cosmos—location, sustainability, security, timing—and the confinement would be excruciating.

Perhaps the prohibitive objection is that you can't get there from here, as the Mainers say. Einstein theorized that the velocity of objects must be less than the speed of light—the "C" in the E=MC^2 equation. In practice spacecraft don't go anywhere near that fast: the velocity of New Horizons, the fastest spacecraft ever built, was—and is—roughly 31,000 miles per hour.[103] Light travels at 186,000 miles per second. The nearest solar system, Alpha Centauri, is 4.37 light-years away from us. If New Horizons were going there, it would take almost 95,000 years.

To go, boldly or not, to a more local destination, say on Mars, would take about a year, followed by a stay characterized by different gravity, artificial air, a limited diet, limited recreation, limited exercise, always the same few companions, and the uninhabitable wasteland outside: none but the most optimistic foresee a long settlement. Plus there's Murphy. On such a venture too many things could go wrong, any one of which would doom the colonists.

Not to mention scale. It is logistically unthinkable that a craft could be constructed in space—and physically impossible that it could be launched from Earth—big enough to carry a significant human contingent. (So too the problem of orbiting endlessly with a societal-sized population.) Space is infinite and infinitely expensive. The world's dwindling resources would never permit the mounting of such a project.

Even if a few human beings could be launched on a colonizing mission, why should they? What good will it do the human population on Earth that perhaps some remnant of humanity might survive beyond our planet? What relevance could our human history have when viewed from an alien landscape? It is as much a moral question as a philosophical one: if escape to the stars becomes necessary, then we will have failed in our stewardship here: why then should we attempt to reward ourselves with a home elsewhere? Supposing that an improbable mission succeeded and a small lodgment were made on another habitable planet, is that who we want to be: a life-form that despoils planets and moves on, somebody else's aliens, humans as space locusts? Wouldn't it be more fitting, and more just, to accept our extinction—or, better, employ our great gifts now toward survival in the beautiful home already provided for us?

Technology as Savior

Something like the same logic and the same objections pertain to the idea of being saved by geo-engineering. The promise seems so apt and so fitting: a more primitive technology got us into this mess, advanced technology can get us out of it. But it can't. The Earth is incredibly vast, and the damage wreaked over two centuries by billions of people working their hardest to fulfill themselves and provide for their posterity cannot be repaired by an idea you could write on the back of a napkin. (This seems like a misconception by some environmentalists as well, that if only the magic phrase is found, people will change their way of life simply by wishing to.) None of the ideas for recapturing carbon from the atmosphere or interfering with the feedback systems that accelerate warming, such as the albedo effect,[104] address the underlying problem—our use of *more*—and in some cases these measures might cause or aggravate other problems already in precarious balance. Nor will we wish to have them in our lives. There's evidence of this already: even as the need for renewables has become so widely accepted, many homeowners who call themselves friends of the environment have opposed wind turbines. How will they, or any of us, feel about hundreds of millions of carbon-capture machines around the world drawing power and doing whatever they do to no visible purpose?[105] Or ubiquitous, permanent CO_2 pipelines and storage sites, applying an involved, artificial process to repeal an imbalance caused by ordinary human life?[106] Or perpetually red skies from a sulfur-dioxide umbrella shot into the atmosphere by missiles or spewed by aircraft (at their own cost in carbon)?[107] Again to the recurring objection: if we turn the Earth into a mechanized life-support system, largely denuded of our co-species—even if we could do this—who would want to live here?

A cousin of geo-engineering is bio-engineering, which among its many applications would replace fossil fuels with genetically-modified biofuels. The flora that produce these fuels would be designed to put down deep roots, to last beyond one growing season, and to act as natural carbon sinks over their millions of acres. In this role they would offset the carbon released when their harvest is burned as fuel—in effect, it would be like planting a tree to offset a car rental. Again: so apt, so fitting. But as a careful journalist of rising seas has pointed out, "[A]ccomplishing this would require a massive expansion of agriculture, sweeping changes to the world's energy infrastructure, bold political leadership, and trillions of dollars."[108] These are the same requirements, impossible to meet within the time desired, that have foiled policymakers from the beginning. In this case, a further problem is that biofuels would compete with the agriculture needed for food in a shrinking arable landmass: *eat or fly?* would be the simple choice.

All these subjects should convince us that there is no quick fix for climate change. The water, the fire, the heat, the drought, and the plague are going to come; we are going to suffer; and if our descendants are to survive with anything like normal lives—if in fact any remnant of humanity will survive—we need to *begin* to change our way of life as soon as possible.

We can do it. The new normal is not so abnormal, and the hardships won't be as hard now as later. For many years, despite the dire predictions, we will be able to affect the outcome. Take a look around you at our beautiful world. It's worth saving. We can do it. You can do it. You can save it. You just have to try.

From Our Institutions: Change and Continuity

What kind of landing?

Economists facing recession speak of the "soft landing." You've experienced that: the wheels kiss the ground so easily you hardly know you're earthbound again, and the world beyond the jetway seems more promising. In the "hard landing" a tire blows or a strut collapses, and the passengers exclaim against the suddenly real, though still small, possibility of death or serious injury. Perhaps you've experienced that too.

The kind of landing I'm talking about is the one where the plane winds up beyond the airport fence, a charred shell on which part of a still-cheerful logo may be seen. But everyone gets out safely.

To restate the problem: our home is becoming increasingly hostile to us, and life will be much more expensive, at a time when there will be less to share among more people fewer of whom will be earning incomes.

Here's a ray of hope: to some degree the economic and physical hardship caused by warming will in itself cause a reduction in carbon emissions. We saw this briefly during the Covid-19 pandemic. The jolt we're going to get in our way of life will therefore be good for the environment and good for us too if we can make the necessary changes before the worst is forced upon us.

And here's another: there is still time. For most of us five years is the horizon. In ten years the kids will be off to college. Twenty years ago the smartphone had just been invented. It's amazing how things can change—how we can change—in a relatively short time. The robot that will send you into early retirement isn't a gleam in its father's eye yet. You, we, all of us have the time to radically reduce our carbon footprints, to embrace *less* and adjust to it, before the real misery sets in. And the most hopeful thing is that our lives will be far more meaningful when we stop centering them around our material possessions.

In these divided times it has become a question whether we can do anything together, even something we mostly agree on. History says that people will set aside their differences in the face of a common threat (often ending their cooperation as soon as the threat has passed). The problem with that model here is that many of our fellow citizens see the measures to be taken as threatening in themselves. The U.S. and other nations have raised the acquisition of wealth to a Maslowian need. Climate change, which threatens wealth, is widely viewed, despite the certainty that we have caused it, as an external force, something that cannot be helped. We should "deal with it" somehow, but only if that dealing doesn't cost too much. Voluntarily giving up luxuries—accepting *less* when a) no individual's sacrifice can make a difference to the outcome and b) the outcome is already determined—seems to many people like madness. How can we bring the airplane down safely?

It's possible. But certain things, mostly in ourselves, will have to change, and others, mostly in our institutions, will have to remain the same. This sounds like a corruption of the famous French epigram,[109] but it is true in any language. Radical change with continuity. Is it possible?

Government

In government we badly need both these things. First and foremost, we need government. I hope you have im-

agined with me how difficult the future will be, the im-
possibility of managing global problems, if either national sovereignty or international cooperation breaks
down. The resolve to make changes and endure hardships can only come from the people, the policies and
resources only from a central authority governing with
the people's consent.

The Preamble of the U.S. Constitution gives a good
idea of what government should do.[110] The general mission of a government, as of any institution, is to solve
problems. Our government isn't solving them; not the
climate problem nor the automation problem nor, it may
be argued, any major problem our country has faced
since the Moon landings.[111] As a result, we the governed
have exhausted hope, patience, and finally frustration,
leaving what feels like a permanent cynicism, that the
more things remain the same the more they will never
change. Our cynicism adds to the inertia we must overcome in order to face the future. More than this, it increases that sense of a breakdown in the order of things
which can make us turn to false solutions and ultimately suffer a real breakdown.

We call it gridlock. The Federal government in all
three branches is at war with itself:

The executive branch comes to office with an agenda,
the first item of which is to undo the deeds of its predecessors; for the president and his team there is no such
thing as settled policy. A recent tactic goes beyond spin-
doctoring to tell outright lies and "double down" on
them. More than one administration has practiced the
facile sort of leadership that cannot achieve lasting results.

The legislative branch is paralyzed by partisan doc-
trine, by the desire to avenge past offenses, by the fear
of losing power to persons even more extreme than they,
and by submissive dependence on campaign contribu-
tors who, like customers, are always right.

The judicial branch, besides being riven by partisanship itself, is torn between two interpretations of the Constitution, strict construction and judicial activism.

All three branches are biased toward privilege, thanks to the influence of special interests—both big business and the single-issue advocacy groups. Executive agencies issue rules that favor the wealthy and that delegate much of their oversight to the overseen. The Congressional budget is larded with pork, often secretly at the eleventh hour. PACs and Super PACs exert disproportionate, undemocratic influence on decisions, yet the Supreme Court has ruled that a corporation is a person and that nearly all campaign contributions are free speech and must not be limited, no matter that big money is so influential it essentially buys votes.

The news media, that unofficial fourth branch of government, which travels in the same chariot and whispers in government's ear, "You are not a god," is gridlocked too. Except for a few powerful media companies, ever mindful of the bottom line, whose print and TV operations reach relatively small audiences, news reporting has been conquered by the internet, the *bête noir* of free speech. Newspapers were once called fish-wrappers to suggest their impermanence, but the internet is even less permanent, which means less accountable: something goes viral and then it is taken down; it does its work and then no one can find it, like a payment under the table. Amid the babel of opinion, fact, error, and disinformation, the responsible Fourth Estate has lost its power to scold; the general in his chariot hears only the cheers of his supporters.

With such a combination, democracy is turning into a new kind of mob rule, anti-elite in spirit but actually manipulated by the rich and powerful. Those left out of the bounty, having been made both cynical and resentful, rally for anyone, rich or not, who fires their resentment—as long as the speaker attacks groups even more vulnerable than they in the crudest possible terms, for vulgarity is the new diction of social intercourse. From all these causes the political atmosphere has filled with explosive vapor.

What, then, must we do? The three branches of the government and the media receive variously their inspiration, organization, authority, guidance, and protection from the Constitution. If the Constitution isn't serving its purpose, we must change it; change, in some cases, the specific provisions that are holding us back and change how we interpret its meaning.

Let me take up the interpreting first, since everything else flows from it. *Strict construction* argues that the Constitution means exactly what it says, according to the world of 1787.[112] *Judicial activism* argues that the Constitution is a living document adaptable to changing times. Those nostalgic for simpler days want the former; those who believe that progress is more important than tradition want the latter. Over the life of the Republic, there has always been this tension: Federalists versus Republicans, Whigs versus Democrats, Democrats versus Republicans, and Republicans versus Democrats.

The checks and balances in the Constitution and the background discussion in the Federalist Papers make clear that the founders were wary of threats to stable government. Now, in one of the ironies of history, strict construction, having been harnessed by the rich and powerful to the detriment of problem-solving, poses a threat to the stability it was conceived to protect. I say this because the language of the Constitution is so different from our own that we have lost the ability to understand it. When language is unclear, it may be invoked as justifying almost any point of view, however outmoded. So we argue about a comma in the Second Amendment while overlooking the fact that militias no longer exist and that a one man with "a modern sporting rifle" can kill more people than a Revolutionary War company of musketeers.[113]

To rewrite the highest law of the land in a lasting way will be a significant task, affecting more than a few passages. Practical considerations too suggest aiming high. Throughout our history we have changed the Constitution by amending it. Amendments were fairly frequent earlier—twenty-seven in 231 years—but with the country so divided there have been none recently.[114] Since the usual process has failed, it is time to try something different: a Constitutional Convention to rewrite the document. A living law must emerge from this, flexible enough to take up new questions. The language itself mustn't be thought of as written in stone, and the case law that follows should distinguish between original values and modern necessity. The Constitution should mandate its own review, perhaps every ten years like the Census; however, the country will generally understand even before the first review whether its new charter has met the most important test, to get government problem-solving again.

A word of warning: the convention would itself pose a risk to stability. The opposing sides on some cultural issues are so inflexible and so passionate that the failure of the convention might lead to violent unrest.[115] It is often wisdom not to ask questions that cannot be answered. But for the future existence of humans on the Earth, the questions now being neglected by our government *must* be answered. Whether we like it or not, the choice is the rule of law, obeyed by all, or anarchy.

So what would a Constitutional Convention look like? It must represent everyone. It must do its work free of outside influence (otherwise, as today, it would not be truly representational). It must have a strong mandate to finish. It must allow time for its delegates to find common ground in the proven way, by bargaining and compromise. It must set an example in statesmanship: of different people seeing the humanity in others and opening their hearts and minds to them; of complex and critical thinking capable of addressing both data and theory with common sense; of the free and dispassionate exchange of ideas; and of the fullest uses of modern expressive language to communicate them.

Leadership

Whatever our struggles with the Constitution, government is failing too in presidential leadership. Since the threat of climate change has been public since 1988, I am talking about six administrations, representing both political parties. The six presidents, even those that made the environment a priority, failed to build a national consensus for protecting it. Implicit in the failure are opposite misconceptions of how to lead.

One misconception is that the head of government may rule by fiat (often called "executive orders"). That has never been true, and attempts to lead this way have always failed, as they are failing now. Government depends on consensus. The man regarded as our greatest president, Abraham Lincoln, knew this: during the most dangerous crisis of our history, Lincoln didn't feel empowered to issue the Emancipation Proclamation, limited as it was, until the Union's victory at Antietam had strengthened his position among the electorate.

The other misconception is that prudent heads of government lead from behind, following not guiding public opinion; that they choose "singles not home runs." Franklin Roosevelt often excused himself to Winston Churchill for half-measures to support Britain in her early defiance of the Axis Powers on the grounds that Americans hadn't made up their minds. By other leaders Constitutional safeguards have been cited to justify taking no action at all. When, in 2013, the regime in Syria used chemical weapons against its own people, an act that Barack Obama had said would cross a red line, he decided against a punitive strike ostensibly because Congress hadn't authorized it.

In reality, neither kind of leadership brings change by itself. The most successful leaders have found that a combination of both is required. The work-flow for a new far-reaching national policy, supported by a broad consensus, is iterative; it looks like a scribble of repeated triangular vectors between the leadership and its constituency: up, down, and across. There may be progress, but it won't be linear, and only the most discerning politicians may be able to detect it.[116] Boldness must alternate wisely with patience.

Americans love their freedom of choice and hate to be told what to do. But once they buy into a plan they will die for it, and that support during implementation will save time in the long run, justifying all the work that went into it. Those who advocate for autocracy, for leadership by martial law, are missing the essential strength of our form of government: it isn't forehanded; it isn't agile; but well led it is relentlessly powerful.

A New Economy

Market capitalism has outperformed other economies because it gets the most from *more.* It inspires the individual (except for those at the bottom of the ladder) to seek wealth, and all the efforts of his life—education, employment, family—point toward this goal. But if *less* is the new future, if we must be done with *more,* should we turn to some other system? W.E.B. Du Bois, a long-lived polymath of American life,[117] thought so. When applying for membership in the Communist Party at the age of ninety-three, he said, "Capitalism cannot reform itself; it is doomed to self-destruction. No universal selfishness can bring social good to all."[118]

Perhaps. But if so, communism and the other recent alternatives, socialism and welfare capitalism, have also failed; failed to understand that aspirations are inherently self-interested and that the operation of the market is more efficient than central planning.

66

Work ennobles us, and good work should be rewarded. A person might be willing to accept less for his work if that's all there is; but if the idle are being paid too, he will resent it. Money is a tangible measure of self-importance, and self-importance is among the strongest of human motivations. Whatever economic system comes next, it must take this into this account.

Eventually we will enter an era in which a great many people, perhaps the majority, cannot work, having lost their jobs to machines. Still they must live—must find meaning in their lives. What then? Anticipating that day, one of the 2020 presidential candidates, Andrew Yang, has proposed a Freedom Dividend, a "universal basic income of...[twelve thousand dollars per year] for every American adult over the age of eighteen."[119] I like the second term better than the first, which smacks of b-list advertising. *Universal Basic Income* begins to get at the problem. Combined with adequate healthcare and a competitive K-12 education, it would be the foundation on which to build a life. Those who wanted more would be free to pursue it (subject to the available resources), without resenting the many others who didn't or couldn't.

A way to begin moving toward this goal would be to state it in a U.S. Declaration of Human Rights, perhaps a first product of a Constitutional Convention. Such a declaration and the national discussion leading up to it would answer the question *What does society owe the individual by right and vice versa by responsibility?*

Giving all adults a guaranteed income sounds like a lot of money. In 2020 there were about 263 million Americans eighteen or older.[120] At twelve thousand dollars each, this one program would cost more than $3.1 trillion in the first year. By comparison, the total tax revenue for 2019, the last year as I wrote this for which data was available, was budgeted to be $3.4 trillion.[121] And that has to pay for all government operations.

Such a stipend might be offset for those receiving Social Security and other entitlements and for those enjoying a certain level of income and/or wealth. While the economy is growing, as Yang and others predict it will (if we elect them), the program might be affordable, a return to the welfare state. But as I have been saying, a growing economy isn't—and shouldn't be—in our future. Whether we cause it or not, markets will contract and revenue will fall. Some new economy, freely chosen and not imposed, must therefore emerge to meet the basic needs of a population under duress from the climate and idled by machines. What will it look like? How will we afford it? How will the income be distributed? Should higher income be a value? These are deep questions, for whose answers we will have to put our trust in time. But perhaps we can begin now to ask them.

Looking Beyond Profits

Some people are—asking deep questions and looking beyond profits, in unexpected quarters. At the Business Roundtable, the CEOs of 181 companies have issued a "Statement on the Purpose of a Corporation," which raises the importance of other stakeholders—customers, employees, suppliers, and communities—to that of investors.[122] Presumably "communities" includes the world community and its climate. If they mean it, it's a start. Next might be the breaking up of corporations that have become "too big to fail"; we did that with AT&T in the 1980s, under a business-friendly president. Next might be the reimposition of taxes on windfall profits and a much higher upper tax bracket on personal income. Next might be the rise of B Corporations. Next might be new ideas about lending and interest, one of which is that access to capital is a right. Next might be a reordering of national priorities with a view toward the world as it is, not as it was during the Cold War; away from the military-industrial complex and all the other complexes.

From Ourselves: Old Values for New Times

Taking an Interest in One Another

I am at risk of circular reasoning: government has failed to bring us together, therefore we must look to government. Clearly the solution begins—as the problem did—with us: we must learn to respect and listen to each other. I would say "once again," for this isn't the first time we've seen such political passions; in fact American history has had few eras of good feelings, and those were often derided as conformist and stultifying.

But today feels different, even for those of us who lived through Vietnam and the civil rights era, when the divisions often tore families apart. As many observers have noted, we have separated ourselves from casual contact with other people. We don't know what's on their minds. We don't learn their life stories—the basis for their beliefs. The absence of familiarity has bred distrust, fear, and profiling.[123] This hardens, until we tend to hear only what we already believe, and we reflexively disregard contrary opinions—and those who hold them—as ignorant, foolish, and malignant. And unpatriotic. If you voted in the 2016 election, you've had trouble since then seeing your opponents as caring and responsible people who work hard, are devoted to their families, struggle with the same problems, and want from life essentially the same things that you want. And yet they are and they do.

Hardly the basis for coming together to face an existential threat.

Technology plays a role in our alienation, particularly in reducing our attention span, but we faced alienation even before life was wired (or wireless). For hundreds of years much of white America viewed people of color as an inferior race. To change that attitude required the bloodiest war in our history followed by a hundred years of repression, murder, and stalemate on one side and non-violence, faith, and patience on the other.[124] The fear of the black man began with the legend of his carnality, from which no Southern woman was thought to be safe, although in point of fact actual miscegenation occurred mostly when white masters impregnated black slaves. With radio and television we began to see past the stereotypes—began to feel the connectedness of all lives beneath the skin. Here is another irony, that the machines that have helped us despoil the planet and increase our isolation from each other have also helped us get along better, for a time.

We must find a way to love our neighbors. Not because it's a virtue—though of course it is—but because our survival depends on it. More logic:

☞ We can best meet the climate problem by sacrificing our desire for *more.*

☞ Each of us is capable of sacrifice as long as the sacrifice is fairly shared.[125]

☞ To be confident of that fairness requires a strong, effective national government.

☞ To have such a government, our laws must evolve.

☞ To pass new laws, we must understand and engage with our fellow citizens.

Although the risk of a breakdown is present, the effort to redress in one setting whatever is obsolete in the Constitution may actually bring us together. A Constitutional Convention, besides publishing a Declaration of Rights

and rewriting our supreme law, would offer the opportunity for truth and reconciliation. For example, what is
the best way to address the legacies of the enslavement
of African-Americans and the genocide of Native Americans? Is the right to bear arms unlimited when guns kill
so many? Since abortion is horrible but people will keep
having sex, and since a growing population itself poses a
threat to our survival, what pro-life principles can be
translated into practical, human measures from which
women can choose? When technology promulgates
falsehoods and images of violence and perverted sexual-
ity, what is the proper scope for freedom of speech?

I am convinced that sincere and open-minded people,
who see their colleagues as also sincere and open-
minded, would be able to come together and decide these
questions and write, clearly, precisely, comprehensively,
and perhaps in timeless language once again, their find-
ings into law.

I will go so far as to offer a recipe card for each delegate:

How to Talk to Your Neighbor

1. Tell your life story. Listen with interest, patience,
    and sympathy to theirs.

2. Take the plunge. Bring up difficult subjects—
    difficult for both of you.

3. Argue fairly, factually, and respectfully, using your
    best words.

4. Share a meal.

5. Repeat.

That might be a card each of us should carry.

Rediscovering that Life Is Funny

None of this will be possible without some basic degree of tolerance, either natural or imposed. Natural is easier. All of us would like to think we are tolerant. Take away what we fear, and we would be more inclined to act that way. With that in mind, I would recommend to the delegates the fullest use of their senses of humor. Humor: the universal icebreaker, the great equalizer, the best medicine. In truth we are all ridiculous in some way. Even the most illustrious of us lose our dignity sometimes. Nothing brings people together like a good laugh.

Unfortunately, our culture has become so wary of giving offense and so quick to take it that, when not shouting at each other, we speak to each other as yoga teachers do. Woe to the public figure who tries to be witty; in fact, all humor except self-deprecating humor is proscribed in public life, and the only acceptable tone on the stump or in social media is a kind of impassioned earnestness.[126] What has become of us? Humor is key to resilience, of which we are going to need quadrillions of barrels before this century is over. Anything can be funny, and no subject should be off limits. In 1939 Dorothy's Kansas house fell on top of the Wicked Witch of the East, whose black- and white-striped stockings and bright red slippers we see jutting out. The hero of *Life is Beautiful* (1997), a Jew swept up in the Italian Holocaust, winks at his son as he marches to his execution. Either burlesque would offend someone today, and as the corporate world has taught us, offended people don't buy products. Like other originally worthwhile things, political correctness has become too successful, too powerful; especially on campus, where it adjures safe spaces and trigger warnings. As if anywhere on Earth is truly safe; as if triggering strong reactions isn't educational. We've got to stop taking ourselves so seriously.

Overcoming Fear With Courage

And we've got to stop being so afraid. Here I come—
"Finally!" you say—to my chief concern, having already
proved, I hope, its ill effects. How do we learn to face the
future with courage?

Again, we examine history. When Franklin Roosevelt, in his first inaugural address, said "We have nothing to fear but fear itself," he was using hyperbole.
There was actually a great deal to fear in 1933, including insurrection by the dispossessed. His four terms in
office, the last mostly posthumous, saw not one but two
waves of economic failure, Pearl Harbor, sixty million
war dead including the Holocaust victims, the Bomb,
and the Red Scare.

The coincidence of these last two was especially damaging. We had been afraid of Communists since the
1920s, but when they got the Bomb, over which we had
hoped to preserve a monopoly, every newspaper headline seemed fateful. "Better Red than Dead" wasn't important so much because of its argument (the converse
was also a slogan) as because that *was* the choice, or
seemed to be. We were going to be enslaved by an evil
ideology imposed from within, by treason...or we were
all going up in a mushroom cloud.

In the 50s and 60s we were afraid even when heady
with success. The domino theory would never have tak-
en root without our fear—perhaps our need—of enemies. Losing in Vietnam, which was inevitable once we
had committed to fight, blighted our courage for a generation. We had one Morning in America during which
the sun intermittently shone, and then came 9/11, as
painful a loss as Pearl Harbor but without the redemp-
tion of victory. As we see now, terrorism has preyed on
our fears more than either world war or cold war. Despite an all-powerful security apparatus (with its con-
comitant risk of leaks), we have been afraid ever since
the towers fell: not in a new "scare" but through pessi-

73

mism, wariness, and self-absorption with no end in sight.

It infects everything we are and do. We cling to technology for the certainty it provides. We track our children by their phones. In our adult relationships we walk on eggshells, afraid of giving offense, unwilling therefore to say what we think, the same reason we suppress our senses of humor. In our fear of litigation we paper the world with legal disclaimers (a manufacturer's sticker on my bicycle advises: *Neon colors may fade.)* Women are afraid of men, and men are afraid of being feared by women. Whites are afraid of blacks, and blacks, with more justification, are afraid of whites. We obsess over the symptoms of our health without really attending to our health itself. (And since we are afraid of seeming old-fashioned, the term *health* has been jettisoned for *wellness.)* A skipped heartbeat sends us to the doctor, who, fearful of not doing enough, prescribes unnecessary tests and/or unnecessary drugs.[127]

As far as the climate is concerned, the diagnosis is in, and now we know what to be afraid of, if something else doesn't kill us first.

The question is, How can we live this way?

My answer is, With courage.

And because I myself am afraid of unjust criticism, let me hasten to add that I'm not being glib here. Courage is a trait, a habit, and a state of mind. Given sufficient motive, anyone can be courageous—in fact almost all of us are at some point in our lives. The soldier who falls on a hand grenade to save his buddies shows remarkable courage, but his fiancée back home shows courage too through months or years of grieving one hour at a time. Endurance is courage, and so is selflessness.

I'm not sure how many born heroes there are in the world. My own courage, such as it is, was learned. The first time I heard a naval gun fired—standing safely about thirty feet from the muzzle—it felt that the ship must have blown up. How could anyone bear such a con-cussion? But the sailors around me didn't seem con-cerned about the firing of the shot—or they were putting a good face on it—and with time I could drink my coffee without spilling it while the gun fired and the bridge windows rattled. As I get older I find myself more fear-ful. But even now, when presented with danger, if I have a moment to think about it, I can remind myself how I responded in the past; I can don the heroic cape again; I can steel myself for what lies ahead.

And because I can, so can others; in fact, even affect-ed courage is contagious.

The courage to meet the dread of climate change lies within all of us, but it isn't just the courage not to run from the idea of the end of the world. It's the courage to make positive changes in our lives with a view, ulti-mately, toward helping to mitigate the problem. There it sits on the horizon. But each of us can act, hoping that the aggregate result will ensure life for our descendants. This is what so many of the experts miss, including those of the highest authority. The world is not going to achieve its carbon reduction goals by 2030, as the U.N. warns it must. The fact that every projection of warm-ing and its effects has proved optimistic should convince us that we must do what we can and live with the con-sequences: periodic disasters and chronic hardships. Courage will allow us to endure both.

So how should you and I think courageously about the climate problem? First, let's not add guilt to our burdens. None of us is a villain—our inherited way of life caused this—and each of us is trying his best. Second, be pre-pared. Every child is going to face greater challenges than his parents, but he can be well prepared to meet them. Third, trust in the odds. The chances of our dying in a hurricane, wildfire, flood, or drought are small; the chances of our starving or dying of thirst are practically nonexistent. Others in the world will suffer more than we, and we must help them. Our own personal challenges will likely be the difficulty of embracing *less,* the financial pressures of lost income and greater expenses, and the hardships of living in a decaying society.

Fourth, be grateful for our lives. Even under stress, our planet is a beautiful place, and life here is a miraculous gift, a vastly luckier chance than any bad luck we may endure while we are here. We can open ourselves to joy and beauty—by conscious act. Whatever warming brings, there will continue to be splendid days, pure and sunlit; we should go outside and enjoy them.

Finally, let us be examples of courage. When, during the Cuban Missile Crisis, I wanted to turn our basement into a fallout shelter, my mother said, "You cannot live your life in fear." At the time I took that as an admonition; now, so many years and hazards overcome later, I see that she was simply stating a fact—she, who was afraid of so much herself. We cannot live our lives in fear.

Telling the Truth

How odd to say that we must tell each other the truth. It used to be that we were raised to do this, even to take pride in confessing: George Washington and the cherry tree, etc. Now we seem to have become enamored of a false value that the admirable thing is to win at all costs, and if by guile then so much the better. We are not ashamed of our lies, we are proud of them. That has to stop. "All's fair in love and war," the winner crows: but it isn't, there are rules for both. The lesson of truthfulness has to be taught from infancy, reinforced through the years of education, and insisted upon in adulthood. The liar should be made to feel his shame. Except on rare occasions, no end is justified by lying. Rather, lying wastes time and effort and goodwill in determining action. The truth will come out eventually, and whatever the liar has won by his lie will be lost, with interest. It's one thing to be clever. It's one thing to be "strategic." But now, of all moments in human history, is the time to be honest with each other.

Curbing Violence

We are a competitive species; nature made us that way. In our history competition—including the competition of ideas—has often meant violence, sometimes on a massive scale. It is a question whether we are more or less violent than we used to be. The urge to kill may live within each of us, but so does the desire to be kind, to be generous.

Today it is violence that makes news and sells video games and fills movie theaters. Perhaps this is because actual violence in our world is declining, so that its representation on the screen carries an allure of the remote and forbidden. Having been exposed to dramatized violence since childhood, with so many images of killing, of conflicts settled and dilemmas solved by it, we are becoming inured to it in that form. In response, the producers of violent programming increase the grotesquerie without thought to the thing being represented. The deaths of actual human beings are presented as merely the last formal movement of a balletic encounter in which the prowess of warriors or the lethality of weapons is featured. Or both. It's all make-believe. We know the characters don't actually die, therefore they weren't real to begin with, therefore no harm is done.

Except that great harm *is* done because we have been conditioned to believe it's a violent world, more violent than it really is, so that when we hear of mass murders we both grieve and shrug. And when we encounter people with ideas contrary to ours, whom we immediately assume to be enemies, our culture has provided us with a template for how to address them. Not with kindness and generosity, as we ourselves would want to be shown, but with violence. We rough them up at political rallies, we send them death threats through social me-

77

dia, we clean our "modern sporting rifles" and say to them in our minds: *Just try that here.*

And sometimes, too often, we kill them.

Because we must act together, changing this way of thinking is a fundamental part of dealing with the climate problem. We humans have an instinct to fight. But history has shown that our powers of reason, curbing baser instincts, can lead us to respond more constructively, for our own good. If we cannot love our neighbors as ourselves—the axiom common to almost every religion and ideology—we can address them with civility, without threat. And then we can begin to understand them.

Once we do, we will be more inclined to share with them, even from our scarcity.

Accepting Uncertainty

Probably the most important thing we can do for ourselves is learn to accept uncertainty. However civilization develops, life will always be uncertain: living with climate change will feel as powerless as the dread days of feudalism. We aren't ready; in our comfortable lives now we are afraid of the untoward and the inconvenient. When we make an appointment, we confirm and reconfirm it, going so far as to text each other of our progress toward the rendezvous. In big box stores husbands text wives to learn which aisle they are in. Subway stations say to the minute when the next train is coming. Road signs say how long to the next intersection. Almost everybody follows automated directions instead of maps—"Turn right in fifty feet" feels like the epitome of precision (as it will until cars drive themselves, when the car's robot-driver will never tire of answering the question "Are we there yet?").

We take uncertainty as a threat, but for many people it adds to the zest of living. Try this: make an appointment, arrive fifteen minutes before the scheduled time, and turn off your phone. Just sit there, waiting. When people around you, who don't like the uncertainty of *you,* ask if they can help, reply that you are waiting for a friend. What do you feel? At first, you probably feel a little anxious. If your friend is late, you may feel disappointed and even irritated. If a no-show (because you didn't confirm!), now what? *That* is precisely where the fun comes in: *now what?* Now you can do whatever you want, whatever pleasurable unstructured thing. You're going to be fine. Life has liberated you, given you a small adventure. Receive the gift. Engage the people who questioned you. Open yourself to the moment. Exercise your curiosity. Do something spontaneous. Embrace the uncertainty.

If you are light on your feet, you will enjoy yourself more, and you will be better prepared for the uncertain future. It's not unthinkable that that future may veer far from your intentions. *You can't take it with you,* people used to say. As far as material wealth that's true, but what you can take with you, and enjoy while you are here, is the greater wealth of rich experiences, good deeds, and love given and received. Let life surprise you.

Practicing Patience

If we can learn uncertainty, we can learn patience. "I *am* patient," you might reply, but you aren't really: not as patient as you will have to be. The future will require of us the kind of patience that lasts for years. How often do you check your phone? According to one study, people check them an average of 2,617 times a day,[128] which comes to being patient for about thirty seconds. We expect things to happen immediately—if not sooner!—and to be at something of a loss if they don't. Is that you? No doubt, when the times demand it, you will adapt; but you can practice adapting now.

The Navy taught me patience, for which I am grateful. I wasn't born that way—quite the opposite. But when I had to sail three weeks across the Pacific to reach a rendezvous for which my ship must not be ten minutes late;[129] or stand, alertly, a four-hour midwatch when nothing was going on; or endure for several days the force of a gale there was no escaping; or stay up all night fixing the ship's navigational position every three minutes off a hostile shore, I learned to live in the present and not get ahead of myself.

That's the biggest hurdle. If my own experience is true for you, you will find at first that your mind is restless. Patience requires patience to make antibodies. You know this if you have ever gotten away to the beach from a high-pressure job. The first few days felt weird, dislocated; and it took some large part of the vacation to prepare yourself to enjoy the rest of it. How relaxed you were at the end!—only to be swept up into your hectic life again almost before the airplane touched down. The idea is to change your life so that mind-time, body-time, and actual time are all the same. *Then* you can be patient, for it won't feel that you must always be off and doing the next thing.

It's a virtuous circle: as you embrace *less,* patience will give you the strength to bear it; as your life tends toward simplicity, your patience will grow.

Before long patience will become a cherished value. At first you will judge others on their mastery of it, but eventually you will understand what the commitment to living in harmony with the Earth really means: that your own actions, one of eight billion, do matter; but that others may not be as far along yet, and the goal must be not to judge but to open minds.

What Can You Do?

Good For You! Also, Good *For* You

This may be the most important chapter of the book, for reasons that go beyond its details. My purpose so far has been to suggest a change in our aspirations vital to the environment and the survival of our species. You may not have a choice. But if you do, why should you aspire to live with less? How will that enrich your life?

The answer is *immeasurably*. Think of your lifestyle now, so dependent on machines. They are convenient, and some of them are nearly perfect, that state we humans have always strived to achieve. Your life in your phone, for example: is it not well organized, accessible, comprehensive, and beautifully contained? Does it not give you the confidence of near certainty? But in what way does your phone, or your computer, or your fitness bracelet *enrich* your life? Does it make you happier? Give you more time? Help you do better work? Improve the quality of your relationships? Exercise your body? Stimulate your creativity? Or, for whatever reason, have you gotten to a point where your day is a constant problem of time management, you have trouble thinking through ideas without being interrupted, some of your most important relationships are sustained only by clever text messages, you are available to your office 24/7 but most after-work contacts are about trivia, you seldom enjoy yourself as you would like, your head is always over your phone to the point of addiction, your sleep is broken, your meals are irregular and made for you by strangers, and you are putting on weight?

It doesn't have to be that way. Without all that apparatus you will rediscover simple pleasures—an evening with a friend or a book instead of with talking heads or virtual beings. There will be more time to achieve balance, the opposite of bingeing, which in turn will improve your knowledge, your judgment, and your service to others. You will learn again to concentrate your mind, which will reward you with ideas that make you exclaim, with pleasure, "Where did that come from?" You will have greater resilience against adversity and greater hopefulness, which is infectious. And, regardless of your income, you will have more money.

Now for the specifics. Remember that all my suggestions flow from a conviction that the future will be hard and that if we do not change our own lives to prepare for it, the changes will be imposed upon us. You can do some things now and plan to do more when possible. I understand how difficult it is to change the circumstances of your life. Do what you can. There will probably be no soft landing this time, but to survive the crash with your hopes intact would be a considerable achievement.

Unplug

Your life isn't really in your phone. Social networking is not the same as having real relationships. Technology and especially the internet are causing your mind to atrophy. Many of the assertions of cyberspace aren't true. Flat screens are addictive and may harm what you care about just as other addictions do. The world is both a beautiful and dangerous place, and you need to live in it with your head up.

You know these things. Summon the resolution to unplug. After a short time, it will prove to be the easiest and most satisfying change on this list.

Fly Less—Or Not at All

If by waving a magic wand I could take every car and truck off the roads, the Earth would still overheat from aircraft. The airline industry is trying to reduce its carbon footprint, a challenging goal. Reductions in ground operations are possible, and some carbon might be saved by changes in takeoff and landing practices. But the physics of flying—marvelous flying—limit what can be done with emissions, and to reach its modest climate goals, the industry plans to rely on offsets, effective—if they are effective—after saplings become real trees.[130] Above all, the industry remains committed to growth.

Often air travel is elective—faster than other modes but not essential. Among the choices for public transportation it is the most harrowing, so why not take a train or bus instead? Or you might travel by car, with greater freedom and privacy. (Each passenger you bring along would share in the carbon savings.)

If you must fly, choose a direct flight on a lightweight aircraft, pack as little as you can (even a pair of shoes makes a difference[131]), and sit in coach. Putting off your trip until cold weather will save carbon because the jet engines won't have to work so hard. Better yet, put it off until....

Don't Drive So Much

Of course you cannot make this or any of the changes on this list overnight. Unless you live in a city, your way of life depends on the car. Yes, you should walk more. Yes, you should tune up your bike. Yes, you should take the bus, train, trolleys, the subway—whatever your community offers. But the odds are that even if you stop driving your children to hockey practice and ballet lessons, you cannot live without a car.

Very well. But use it less. Join a car-pool. Consult with your friends and neighbors before taking a drive—

is there anything they need? Avoid the rideshare companies, which in their constant patrolling increase total miles traveled and therefore carbon emissions. Remember that electric cars are only as clean as the source of their electricity; know where that next charge is ultimately coming from.

Be creative; work at it. Plan for a future in which cars aren't so important to you. See if you can reduce your family's fleet from three to two, from two to one. Can you give up entirely? You will save truck beds of money.

Lobby for Public Transportation

Once you limit your own driving, you will see what your community needs. Some modes of transportation may be expensive and complicated to build; but anything that uses existing roads, like bus service, can be started quickly. If you already have bus service, does it allow you to bring onboard everything you need? If not, push for buses with fewer seats and more liberal policies.

When people travel overland in efficient numbers, profits don't matter: the service is a public good. So it will be with monorail, high-speed trains like Maglev and Vactrain, and self-driving trolleys in the future. Here is one kind of progress we can be hopeful about, a way of remaining mobile while living in harmony with the Earth.

Still, it will be a long time before we give up our dependence on cars. Like other machines they are marvels of engineering and craftsmanship and often beautiful to behold (the exceptions do stand out). But you will live a richer, more productive, and not greatly inconvenient life by taking the road less traveled—the bus lane. And that will make all the difference.

Move to the City

This is hardly fair. You have worked hard to afford the things that rural or suburban living offers—space, among others. You want to be close to nature. The neighborhoods are quieter and safer. The schools are better. Food is cheaper, and housing might be too. Like billions of people who have come before you, you want land.

Yes, *but.* As the climate deteriorates, small communities will be overstretched to deal with the changes. If you live in an area at risk from natural disaster, even a remote risk, it will be harder to protect your investment. Arable land is retreating before drought and flood. To feed our own population, let alone helping to feed the world, farmers will need to reclaim the suburbs.

On the other hand, the city has the resources—and the stake in its own survival—to protect you and your family and offer you a stimulating, enjoyable life with all the amenities. You will thrive here as you cannot continue to thrive in the suburbs. Come visit, then make plans and push yourself to move when the possibility allows. Plenty of good seats are still available—or will be if you demand them.[132]

Save Money

A survey not long ago by the Federal Reserve revealed that 46 percent of Americans said that if they had a sudden need for four hundred dollars, they couldn't come up with it.[133] We hear the horror stories of retirees unable to live on their 401(k) portfolios. The national savings rate has declined to nearly an all-time low.[134] For many years we Americans and our government too have spent money we didn't have, and now we are feeling the effects.

In the future we're going to need those savings. To pay for the costs of disaster in many regions, prices are going to rise, just as our economy, one way or another, is going to retract. The 1970s called this *stagflation*. What lies ahead of us will make the 70s seem like just a blip. You yourself must have a cushion.

Perhaps you must first get yourself and your family out of debt, which as a practical matter means cutting expenses. That will be hard, but it may not be as hard as you think. Draw up a budget and follow it. Instead of rushing out to buy the latest new thing, make the old thing last; repair, reuse, recycle, repurpose it—or buy its replacement the second-hand stores.

Cut back. You can live without the latest smartphones, expensive apps, and premium cable TV channels. You can walk or bike instead of using the rideshare companies. You probably don't need so big a house. You may once again relate to your sports heroes when they aren't club-hopping multimillionaires and you aren't spending hundreds of dollars on a day at the ballpark. Your child may wish to go to a "name" college, but you can counsel "value"—the sweet freedom of a perfectly good education that doesn't require ten years to pay off, longer than indentured servitude.

Getting out of debt and having adequate savings will make the future easier to handle. Ironically, if you spend less, it will be worse for the economy, but those blows are coming anyway. Be prepared.

Cultivate Your Sense of Beauty

This isn't some touchy-feely value but a practical tool for survival. Assuming that the predictions of an automated world are even roughly accurate, machines will take over most of the work now done by humans.

Earlier I suggested that automation may have its limits. Just as it will be difficult to give machines a human consciousness, so it will be difficult to give them an aesthetic sense, which forms in us as a result of living. We can't all be poets. But even after machines take over, the world can be made more beautiful; and there will always be a market for craftsmen as long as the rest of us desire their work.

For proof of that we need look no further than our modern buildings. They are tall, they gleam, they may be (they should be) LEED-certified to the highest standards. But are they really beautiful? Only if your taste runs to the generic and sterile. By any standard they aren't *crafted*. They have been built with the latest materials assembled in the most efficient and profitable way. Few of them include interesting facades or any visual embellishments because that would be more expensive. Yet the Earth still offers wood, stone, and brick, which men and women have learned to carve into beautiful objects.

Walk around cities that have preserved their nineteenth century buildings; compare these to the latest high-rise—even the wrought-iron railings are beautiful. So it can be in anything made by people. Beauty has life and gives life if only we will look for it.

What Can You Teach Your Children?

First, a Pat on the Back

A message to parents: thank you for your service. You are raising wonderful children, often at a sacrifice of your own needs. I submit the following not as a critique of modern parenting, which has developed apace with our culture, but because the peculiar challenges facing us in the future will call for different skills. The new and largely technical skills will be obvious when required. What I present in this chapter are the timeless skills, at present out of fashion, that we ought to reemphasize. As with everything in this book, these prescriptions may apply to both you and your children; and to their children and so on.

Self-Reliance

Like courage and patience, self-reliance seems to be a moral value, intrinsic to character. As such it is easily misperceived. Because we ourselves have become self-reliant, we believe we are bringing up our children that way. The truth, I submit, is the opposite. At a time when communities are safer than ever, the fear of "stranger danger" together with the desire for certainty causes many parents to know too much about their children's lives.

What Can You Teach Your Children?

Once again the culture makes it easy. America runs on cellphone. Children, almost as soon as they can talk, beg for them, the phones being desirable in themselves and as status symbols. Sooner or later, citing convenience, safety, and peer pressure, their parents give in. Doesn't it start this way? From then on their children are always in touch and seldom alone. At home everything they do is subject to inspection. At school they are under a teacher's care—indeed many are signed in and out like accountable goods. After school, they are driven to activities—the car is another highly supervised environment—where a parent coaches or monitors them. In short, much of their time, including weekends and vacations, is planned for them. Modern childhood is an experience shared with adults.

Other than requiring a lot of effort by the adults, why is this bad? It's bad because children aren't being allowed to make their own mistakes and learn from them; and because each child is encouraged to think that the sun rises and sets on him, with "his team" always hovering in support.

Life will decree otherwise, certainly, but the question is when? The children of climate change will be tested early on. They will need to draw on judgment beyond their years: to sift truth from untruth; to think deeply and with due regard to life's complexities; to relate to others, who also wish to feel important; to take initiative without waiting for a rescuer to swoop in; to respect pain but to rise above it when necessary; to try, to fail, and to try again; and to know how to deal with success, impostor as it might be. All this in a world much harsher and less forgiving than that in which their parents grew up.

Good judgment develops over time from a young age, and parents must seize on its teaching moments. *Give them roots and wings* is the household proverb. As sinking roots becomes ever more difficult, strengthening wings must be emphasized.

Imagination

Besides gaining actual experience, children may practice self-reliance through their imaginations. But the opportunities for that are slight because so little of their time is unstructured. When I was a boy, my parents said, "Be home for dinner"; the rest of the day belonged to me. As I roamed, I daydreamed. Walter Mitty's adventures were no more fantastic than mine—or yours, or your children's when they have a chance. I lived in two worlds, solving the problems of each. Imagined glory healed real-world hurts. Achievement in the real world, slow in coming, ratified my imagined sense of worth. Since then, the world of my mind has been invaluable in anticipating and understanding the world around me and in solving the problems it posed.

Solving them sometimes in unusual ways; the future will require that. Procedure manuals won't take anyone far (another reason not to fear machines). What your children will need for problem-solving is a well-developed sense of metaphor—*A reminds me of B: what if it works in the same, or sort-of the same, way?* Metaphor is simply focused imagination. All of us can learn to use it.

All. A young friend of my wife is a terrific artist. Her work is exquisite. But she complains that she has no imagination. Of course she does!—she just hasn't learned the trick of it, which is to listen to her imagination, to escape into it. She hasn't allowed herself to ask *What if...?* I worry about this with her generation. They aren't getting that opportunity. When the time comes, they will need to be able to count on their inventiveness—on what used to be called, in our pride, *American Know-How,* the product of freedom, imagination, and confidence. All can have it, but they need to try.

Send them out to play. Tell them to be home for dinner.

Language Skills

Honest, self-reliant, and imaginative as they may be, our children will still have to work to understand each other, and here we need to help them improve their literacy. I don't mean test scores; I mean expressing clearly and persuasively their most complicated thoughts and feelings. Our citizens today have far more education than citizens a hundred or two hundred years ago, but read the letters of the soldiers in the Civil War, read the applications for relief by farmers in the dust bowl of the 1930s, and you will find a use of language that we cannot begin to match. Our children are using language at the level of the movie punchline and the 280-character tweet; and with emojis and camera shots it's going to get worse.

I remember a generation ago college teachers were complaining that their first-year students didn't read and write well enough to do the work. Those students have become the teachers of today. *Their* students are even more challenged. Who among our children can diagram a sentence? Who cares about the etymology—that is, the origin—of words? Who listens for connotation? But those things matter. Life is far more complicated than it was in the dust bowl. Our children's language must be sophisticated enough to express nuance, which unties complication. Their diction must be precise enough to convey their deepest meanings, uninflated ("great" doesn't mean "good") and undiluted (a million "likes" doesn't make something truly likeable). Only by mastering syntax can they convey those meanings clearly and with persuasive force. Technology hasn't helped, any more than it helps cultivate our minds generally, but the problem started long before the first word processors.

This isn't an old man's cavil: literate use of language is fundamental to cooperation in solving problems. It can also bring pleasure. Let your children read difficult books; let them listen to people who speak well (you! at the dinner table); let them try out their diction and syntax, extravagantly, ridiculously. They will open themselves to new ideas and realize that few truths are absolute and that no one has all the answers. When frustrated, they'll be able to describe their feelings. They will empathize with others. They will begin to understand the world they've been commissioned to save.

Respect for the Law

We have always been fascinated by criminals, iconoclasts, and rebels. Lawyers and judges we see as obstructionists or worse. When the Arab Spring began, we welcomed it, dismissing the overthrown regimes as autocratic and corrupt, so much so that we put in jeopardy our international standing as reliable allies and resolute adversaries.

But the law, with all its obstruction and delay, is what keeps us from anarchy, and as the Earth becomes harder to live on, the law may help keep us from destroying each other. We cannot face climate change without a stable government—if I haven't convinced you of that, please return to page 20. To have a stable government requires that all of us obey the law, whether we agree with it or not.

This is a value we must teach our children, in opposition to the message they are receiving from the culture that glamorizes lawlessness just as it does violence. Children grow up wanting to do what is right. They still have to learn that what is right is relative, open to different interpretations, while the law is specific and concrete, as specific and concrete as good writing and learned interpretation by the courts can make it.

You say you want a revolution? You don't.

Homesteading Skills

Even if the institutions survive and our children are able to live much as we do, they will be more confident if they know how to be pioneers. Each child should be taught both to grow food and to forage for it in the wilderness; to build and furnish his own shelter; to make clothing from raw materials; to perform basic medical tasks; to repair plumbing, electrical systems, and machines; and to manage livestock, particularly horses. Any of these things may be needed in even the easiest future life.[135] As a bonus, learning them will be fun.

You may be asking *Is this really necessary? Pioneers? Horses? Wouldn't making their own clothes assume that our institutions have failed?* I think the answer lies in the eighty years, and this, in fact, is the crux of my book. For the most part, however accompanied by singular events, climate and automation change will be gradual, from month to month imperceptible. But over time our children and their children will come to a new relationship with each other, with the Earth, and with their institutions.

They will be living in an era of less. Growing and preparing their own food will probably be the first of a number of sensible steps to take. Spinning, weaving, sewing, and knitting will make a comeback because they give pleasure and because there will be time for them. If our descendants rediscover the old, as they should, they will need to maintain it, however well-made it is (even the Maytag repairman gets called occasionally). They may decide to work with their hands at non-repetitive tasks that machines can't do or can't do well: as plumbers, carpenters, electricians, and, we hope, bricklayers and stonemasons. With such skills and when money is tight, they will look after their homes.

In this way the future won't be just a rolling back of progress but an amalgam of the modern with the best practices from the past: robots to do much of the work;

public utilities, public safety, public health, and public transportation; a limited system of banking and finance; but also, by degree, what is homemade, home-repaired, and bartered. If there will still be cars on the streets, there will be even more bicycles and, I suspect, not a few horses.

Worldliness

Until now use of this term has generally meant so-phisticated, street-smart, jaded—the opposite of inno-cent. We should restore its literal meaning and teach it to our children.

They will need to know and love the natural world. Plants and animals, rocks and rivers, earth, ocean, and sky: all these things, which are nothing like the world as represented on their phones, must be appreciated in their variety and irregularity. It is not only humans they will try to save.

However, they will also need to know and love the human world. I keep saying its survival will depend on cooperation; let me add now, cooperation with people different from our children. This will require them to learn foreign languages, cultures, history, and geogra-phy. As with the natural world, Google cannot help them, no matter what it claims: they will need to dis-cover the human world for themselves. These children of a country famous for its insularity must travel.

So let them travel. Let them use their self-reliance, imagination, literacy, respect for the law, and pioneer-ing to see the world with the smallest possible carbon footprint. Volunteers join the Peace Corps believing that they will share their advanced skills with others. The mission of the Corps acknowledges something of that, but more important by far is the exposure of people in the host countries to what is best in American life *and vice versa.* Let our children experience the *vice versa.*

What Can You Teach Your Children?

Never before has humanity had more in common or needed it.

Will they want to learn these things? More generally, as they grow up, will they be ready to face the future? Here again I recur to the problem of fear and the necessity for courage. Until the worst, which may never happen, they will continue to find much to live for—love and joy and beauty and wonder. And occupation. I say again, the changes will be slow, although too fast for comfort. With the help of others, with kindness, generosity, and tolerance, they will make the best of whatever bounty the Earth and our civilization may yet provide to them.

It should be enough. There will be loss and grieving, as there is in every life. Ultimately our children will return their own small carbon store to the Earth, as you and I will—before them, we hope. Time will leave us all behind. While we are here we mean something, in ourselves and to each other.

What Do We Owe Each Other?

Fair Play

It is a truth universally acknowledged that the willingness to sacrifice requires that the pain be shared. As Americans we are ever on the lookout for cheating. We resent billionaires who don't pay taxes. We hate it when the administrators of charities go on junkets. The idea that someone could buy an organ transplant feels wrong to us. When celebrities bribe college officials to admit their children, we applaud their being sent to jail. Passports and TSA prechecks seem purposely expensive, as if the poor don't need to travel. I myself think you don't deserve an HOV lane if you're the only O in your V.

Yet since the Reagan Administration—a glittering time, which, let us not forget, ushered in the greatest release of carbon into the atmosphere—the trend is to reward the rich for the advantages they supposedly confer upon society, like the housekeepers who must surely exult when their employers buy a yacht. As we look toward the future, it is natural to suspect how the pain will be shared. We are troubled by the possibility that the rich will be first in the lifeboats and the poor will go down with the ship.

The possibility is especially real in the U.S., where the gap between rich and poor is historically wide.[136] Around the world the concern about fairness varies according to wealth, national power, moral authority, carbon footprint, and vulnerability to the consequences. Under the aegis of the United Nations, nearly two hundred member states with their diverse points of view are trying to work out what the wealthiest, who by no coincidence have contributed most to the greenhouse effect, owe to everyone else—some of whom don't even have a lifejacket.

The Problem of Responding

Four types of responses to the crisis have been defined by the U.N., of which the first two have received the most attention so far:

1. Mitigation, reducing the emissions that cause climate change;

2. Adaptation, improving the ability of nations to deal with its effects;

3. Financial assistance for loss and damage when adaptation fails;

4. Technology development and transfer; for example, geo- and bio-engineering.

With mitigation, the Paris Agreement aims to have each nation's emissions peak as soon as possible, so that the goal of net zero (sources matched by sinks) may be achieved by the year 2050. To that end the Agreement allows nations to trade carbon credits. This is the back half of a "cap-and-trade" policy, by which a legal, mandatory cap is set and then participants may trade credits to stay under it. There is something unfair about it, like getting the nerdy kid to write your poem for you in English class, but it works to reduce total emissions. But the signers of the Agreement deliberately avoided caps; they chose voluntary goals, "Nationally Determined Contributions" (NDC), instead, and there is no enforcement. "Name and encourage," said a U.N. official, avoiding even such coerciveness as "name and shame."[137]

Adaptation is getting particular focus just now,[138] which doesn't speak well for the hopes of the mitigators (but does reflect the ever-more-alarming data from climatologists). This and the Loss and Damage initiatives from the third part of the U.N. program have the potential to break the bank. Imagine all the coastline around the world that someone will want protected from sea and storm; the island nations and low-lying nations under the threat of inundation; the heat, drought, and fire that will drive millions from their homes; the countless emergency rescue, recovery, and relief operations; the industries gutted, the economies crashed. The expenses of adaptation will be staggering. With the best will in the world— the desire by the rich nations not to have a flood of environmental immigrants on their borders—physically it cannot be done.

The Problem of Money

If something is both expensive and cannot be done, hardly anyone would want to pay for it, and this is one of the two things impeding climate payments by the rich nations to the poor. The other is human nature, the fact that charitable feelings are short-lived while self-justification is endless. Mr. John Dashwood in Jane Austen's *Sense and Sensibility* promised his father on his father's deathbed that he would take care of his stepmother and three half-sisters, thinking to give each sister a thousand pounds sterling; in one conversation with his worldly wife ("worldly" in the usual sense), the gift was successively reduced to nothing.[139] We can justify any behavior that accords with our self-interest.

In determining who should pay for climate change and how they should pay and how much, the polite language of diplomacy has occasionally given way to rude accusations,[140] but within the U.N. climate framework a financial modus vivendi has been worked out with wide support (a good example of why in the worst of crises institutions matter). For mitigation and adaptation, a Green Climate Fund has been set up, with its headquarters in Incheon, South Korea, to handle the $100 billion per year committed in principle by the developed nations for the period 2020-2025. As of July 2023 $9.31 billion had been pledged, most of which has been earmarked for the costs of starting the fund.[141]

If this foreshadows extremely high administrative costs, the fund will fail (a good example of why institutions have limits), but it may fail anyway if the rich don't keep their commitments. The developing nations see climate assistance as a moral obligation, but that won't matter to the payers; like the Dashwoods they will talk themselves down to a lesser amount. Nothing in the Paris Agreement, neither NDCs nor the financial flows, is legally binding. Even if it were, the mood in many countries is to defy international organizations, particularly when they ask for money.

Another impediment to the Green Climate Fund is bureaucratic. Recall that the sea barriers in Venice took fifty-four years to build. That was a local initiative, overseen and supported by Italians. If, after achieving consensus, Venice had had to go all the way to South Korea for financing and approval, the red tape would have added years to the project and millions to the cost. In the same way the awkward combination of distance and micromanagement is one of the reasons why the Brexit supporters wanted to Leave the EU. Bureaucrats don't like delays any more than the rest of us, but being afraid of making mistakes—and in this case on an enormous scale, with horrific consequences—they are naturally cautious and meticulous. It's no wonder that most of the $9.31 billion pledged so far has gone to startup costs: the managers of the fund are trying to do it right. But for poor countries that can't wait, the delays could be fatal.

So the climate problem will keep getting worse, cultures and languages will disappear, and the moral obligation of the wealthy nations to return some of the profits of carbon to the poorer nations will not be met, at least not in full. What's to be done?

Money by Another Name

One form of compensation would allow some nations higher NDCs so they might catch up economically. To the credit of all the signers, particularly those whose economies would have benefitted the most, this idea was rejected at Paris. Fair or not, it would have been suicidal. The climate cannot withstand the unchecked emissions of a China or India—as the Chinese and Indians have been reminded, if they needed a reminder, by the noxious brown clouds in Beijing and Delhi—or by the African states, whose populations are projected to quadruple in this century.

Once again we confront reality: the collision has occurred, and the ship is listing and filled with smoke; the survival of the passengers and crew matters more than fairness or accountability. As a practical consideration, none of those who profited or suffered from the carbon inequity will be around for a reckoning. None of those alive in the next century, whether prosperous or doomed, will know their names. Nor will they care how the world got to that point, only that it did.

The Rich Must Meet Their Obligations

It is fair, however, and realistic and wise for the rich nations to be prompt in achieving their NDCs. Although the size of their economies will keep them from carbon neutrality as soon as the developing nations—again, so much *more* cannot be replaced overnight—with their greater resources they should be especially active in working toward that goal. Some are. The others, who are falling short of their promises, like the U.S. and

101

about half the G20 states,[142] having had years of "name and encourage," should now be subject to penalties.

Penalties will carry consequences. Developing nations may hesitate before imposing them, but the need is paramount, and in an interdependent world, when the Commons has been overgrazed, natural resources carry a power of their own. For example, the world's major exporters of lithium, led by Australia, Chile, and China, might refuse to sell it to defaulters; within a short time high-tech would be in crisis.[143] Egypt, Panama, and Turkey might deny the use of their waterways, confounding the defaulters' trade. On a symbolic scale, Nepal might do the same thing with licenses to climb Everest. Among unnatural resources divestment works too. Almost every nation has leverage.

The Value Beyond Value

So far in this chapter what is owed has been money—or NDCs treated as money since they represent economic activity. Money is the most liquid, flexible form of assistance; even when it is earmarked, the recipient has some discretion. As valuable as it is, however, it is not the most valuable: the fundamental responsibility of the rich to the poor, which the rich must meet under penalty of losing all they have, is to protect everyone's right to live.

Climate change will cause millions of refugees. To some of these, such as the Marshall Islanders if the seas continue to rise, we in this country are legally obliged to offer a home. We should expand that obligation to take up our fair share of *all* refugees anywhere in the world. If determined by our proportion of global wealth, our fair share of refugees would be about twenty-five percent.[144] If by proportion of carbon emissions, fourteen percent.[145] If by population, four percent.[146] Without apologies, then, for our wealth, which Americans have worked hard to achieve, and with only a modest increase in population, accepting immigrants based on our carbon emissions would not only be just, it would be the reasonable middle. If two hundred million people were forced to abandon their countries by 2050, as the U.N. predicts,[147] our fair share would be twenty-eight million, or about twice the number of undocumented persons whose futures we are having so much trouble deciding today. Saving the lives of the twenty-eight million and giving them homes would increase our population density by eight persons per square mile.[148]

A large immigration will change forever the character of the United States. But that change is already underway, and we ought to continue it out of self-respect if nothing else. We cannot justify selfishness in this crisis. What it means to be an American is different and will be different from now on, providing that humans survive. We must accept that. As citizens of the world, as citizens of the greatest, richest, luckiest country in the history of the world, let us protect the right to live for everyone, as a value. Isn't that what we think of when see Lady Liberty—liberty *and life?*

A Noble Sentiment: How Do We Get There?

As I said before, important national decisions require an iterative process of grassroots activism and governmental leadership. This might begin with a U.S. Declaration of Human Rights. For more than two hundred years, we have based our values on the Declaration of Independence, whose language moves us still. But in the progress we have made since then, a document justifying the renunciation by thirteen coastal colonies of submission to their overseas king is no longer relevant or complete: why shouldn't we, with all the good writers we have, compose something for the *next* two hundred years, as far as we can foresee them; give the National Archives a new document just as beautiful to put under the glass; something for all of us to live by? Such a Dec-

103

laration would be the obvious first product of a Constitutional Convention; having restated our rights, the delegates would then look toward securing them.

To bring back the greatness of America, we must start doing great things again.

Less Is Enough

Save the Earth. Don't blow ourselves up. Limit the size of families. Renew the Commons. Control machines lest, in the worse case, they control us. Preserve the world order of governments and economies, however the institutions themselves must change. Above all, be realistic.

And if I too am realistic, I must admit that it won't look exactly like my prescriptions or like anyone else's either. The future will determine itself, as it always has. Something new out there will bend—or snap—the arc of history: a breakthrough in technology, a catastrophe no one prepared for. This may be why we are so fearful; overwhelmed by scientific research, we expect the worse, we *intuit* the End of Days. Our civilization feeds on pessimism now. Our favorite art-form, across all media, is dystopic.

After many revisions and a lot of soul-searching, I believe I have stated in general what the Age of (...) may bring. Unfortunately, it makes part—the most important part—of the title of this book: *Less.* The increasingly hot future will be metaphorically covered by a wintry grey sky (if not a sulfur-red one). Life will be modern, as usual, but with new primitive elements. Some of our prospects will be bleak; all will be changed. We will have to learn to take only what we need, to build well and beautifully, and to leave behind enough for our successors. We will be in a more intimate relation to the Earth, even as we receive less bounty from it. While we correct the errors of centuries of stewardship, we will need to withstand many hardships and blows from the climate, and we will be under pressure to respond in our usual way, with greed, selfishness, and competition. Prophets and philosophers have taught us better, to empathize with others, to live by the Golden Rule. To be honorable and just. To be courageous and provident. To laugh. The Earth will be our home for as long as we may keep it. There is no other home for us. The truth is that no matter how much we accumulate and how safe we try to make ourselves, any of us may become a refugee.

Conclusion: Just Start

Do you think I've been extreme? That's not my nature. In my defense I will say that I am trying to look ahead eighty years. Imagine the predictions in 1920 about 2000: who would have thought?

The point is, today we seem to be paralyzed by fear and other venoms, and instead of thinking of 2030 or 2050 as a deadline, which implies a binary outcome, live or die, we should think about tomorrow, by which I actually mean *tomorrow.* Our problem is not that we have failed to surmount this nearly insurmountable challenge: it is that we haven't properly started. None of us can surmount it by ourselves, nor by any single action. But eight billion of us, doing what we can, changing more than we would like, offer the best chance that we have. We may not reach net zero. We may not reach one point five. But we should, by any method of reasoning, give the children of the twenty-second century the prospect of saving themselves and millions of other species. If worse comes to worse and the survivors have access to records (including this one), they will know that at least we tried. With our amazing minds, which no machine can equal, I'm betting the future will be happier than that.

About the Author

Thomas Corcoran entered the Navy in 1972 and retired in 1993 as a captain, having served during the Vietnam War and Operations Desert Shield and Desert Storm. He commanded two ships and was selected for command of a destroyer squadron. He lived in Navy ports on both coasts and deployed in many regions of the world. Besides sixteen years of sea duty, he served in the Pentagon as Special Assistant to the Chief of Naval Operations and as Military Assistant to the Secretary of Defense.

Afraid of Less is his first book of nonfiction following five books of fiction.

Notes

ABBREVIATIONS USED IN NOTES:

CLOCK John Mecklin, ed. The Doomsday Clock. Science and Security Board, *Bulletin of the Atomic Scientists,* "A time of unprecedented danger: It is 90 seconds to midnight." TheBulletin.Org/doomsday-clock, January 24, 2023, retrieved August 15, 2023.

IPCCAR5 IPCC, 2013: Summary for Policymakers. In: *Climate Change 2013: The Physical Science Basis. Contribution of Working Group I to the Fifth Assessment Report of the Intergovernmental Panel on Climate Change,* T.F. Stocker, D. Qin, G.-K. Plattner, M. Tignor, S.K. Allen, J. Boschung, A. Nauels, Y. Xia, V. Bex and P.M. Midgley, eds. Cambridge, U.K. and New York, NY: Cambridge University Press.

IPCC1.5 IPCC, 2018: Summary for Policymakers. In: *Global Warming of 1.5° C. An IPCC Special Report on the impacts of global warming of 1.5° C above pre-industrial levels and related global greenhouse gas emission pathways, in the context of strengthening the global response to the threat of climate change, sustainable development, and efforts to eradicate poverty,* V. Masson-Delmotte, P. Zhai, H.-O. Pörtner, D. Roberts, J. Skea, P.R. Shukla, A. Prani, W. Moufouma-Okia, C. Péan, R. Pidcock, S. Connors, J.B.R. Matthews, Y. Chen, X. Zhou, M.I. Gomis,

E. Lannoy, T. Maycock, M. Tignor and T. Waterfields, eds. In Press.

IPCCAR6 IPCC, 2022: Summary for Policymakers [P.R. Shukla, J. Skea, A. Reisinger, R. Slade, R. Fradera, M. Pathak, A. Al Khourdajie, M. Belkacemi, R. van Diemen, A. Hasija, G. Lisboa, S. Luz, J. Malley, D. McCollum, S. Some, P. Vyas, (eds.)]. In: *Climate Change 2022: Mitigation of Climate Change.* Contribution of Working Group III to the Sixth Assessment Report of the Intergovernmental Panel on Climate Change [P.R. Shukla, J. Skea, R. Slade, A. Al Khourdajie, R. van Diemen, D. McCollum, M. Pathak, S. Some, P. Vyas, R. Fradera, M. Belkacemi, A. Hasija, G. Lisboa, S. Luz, J. Malley, (eds.)]. Cambridge University Press, Cambridge, UK and New York, NY, USA. doi: 10.1017/9781009157926 .001

WARN1 Henry W. Kendall, Union of Concerned Scientists, with 1,574 co-signers, "World Scientists' Warning to Humanity," 1992, uscusa.org, reprinted 1997, retrieved October 4, 2019.

WARN2 William J. Ripple, et. al., "World Scientists' Warning to Humanity: A Second Notice," 2017, *Bioscience,* academic.oup.com/bioscience, retrieved October 4, 2019.

UNEP18 United Nations Environment Programme (UNEP), "Emissions Gap Report 2018," Nairobi: UNEP, 2018. Unenvironment.org/emissionsgap, retrieved October 6, 2019.

Notes

NOTES:

[1] Wikipedia, "Greenhouse gas emissions," August 14, 2023, retrieved August 15, 2023.

[2] As I write this the number of deaths continues to mount from wildfires in the once lush Hawaiian island of Maui.

[3] David Wallace-Wells, *The Uninhabitable Earth: Life After Warming* (New York: Tim Duggan Books, 2019), p. 4.

[4] From now on, for simplicity and unless I need to make a distinction, I will use "carbon" as shorthand for all greenhouse gases, which include carbon dioxide (CO_2), methane (CH_4), nitrous oxide (N_2O), and halocarbons (various elements in which carbon atoms are bonded to one of four types of halogen atoms). IPCCAR5, p. 12.

[5] On June 23, 1988 Dr. James Hansen, director of NASA's Goddard Institute for Space Studies, testifying before the U.S. Senate's Committee on Energy and Natural Resources, was reported as saying:

> Global warming has reached a level such that we can ascribe with a high degree of confidence a cause and effect relationship between the greenhouse effect and observed warming.... It is already happening now. —Philip Shabecoff, "Global Warming Has Begun, Expert Tells Senate," *The New York Times,* June 24, 1988. p. 1.

> The greenhouse effect has been detected and it is changing our climate now.... We already reached the point where the greenhouse effect is important. —Michael Weisskopf, "Scientist Says

Greenhouse Effect is Setting in," *Washington Post,* June 24, 1988.

Hansen's testimony made headlines. He was also co-author of a paper published in Science, on August 28, 1981, but without the same public notice, that linked atmospheric warming with "measured increases in atmospheric carbon dioxide." Wikipedia, "James Hansen," October 4, 2019, retrieved October 14, 2019.

[6] Bill McKibben, *Falter: Has the Human Game Begun to Play Itself Out?* (New York: Henry Holt and Company, 2019), pp. 72-75, 90-101.

[7] WARN1.

[8] David G. Victor, Keigo Akimoto, Yoichi Kaya, Mitsutsune Yamaguchi, Danny Cullenward, and Cameron Hepburn, "Prove Paris was more than paper promises," *Nature,* August 3, 2017, pp. 25-27. And Joeri Rogelj, Michel den Elzen, Niklas Höhne, Taryn Fransen, Hanna Fekete, Harald Winkler, Roberto Schaeffer, Fu Sha, Keywan Riahi, and Maite Meinshausen, "Paris Agreement climate proposals need a boost to keep warming well below 2° C," *Nature,* June 30, 2016, pp. 631-639. Cited in Wikipedia, "Paris Agreement," October 7, 2019, retrieved October 12, 2019.

[9] UNEP18, p. 8.

[10] McKibben, p. 70.

[11] Wikipedia, "Geologic time scale," October 2, 2019, retrieved October 14, 2019.

[12] CLOCK, p.17.

[13] When his 1984 presidential campaign adopted the slogan "It's morning again in America," apparently they had not talked to Dr. Hansen. Wikipedia, "Morning in America," June 18, 2019, retrieved October 14, 2019.

[14] I don't know that the provident pig was male. For singular generic references I use the male personal pronouns, "he," "him," "his," and for the plural "they," "them," "theirs."

[15] Wikipedia, "The Pentagon," October 13, 2019, retrieved October 15, 2019.

[16] American Merchant Marine at War, "Liberty Ship SS Robert E. Peary built in 4 days, 15 hours, 29 minutes," usmm.org, June 5, 2000, retrieved October 15, 2019.

[17] Wallace-Wells, p. 59.

[18] IPCCAR5, p. 10.

[19] Jeff Goodell, *The Water Will Come: Rising Seas, Sinking Cities, and the Remaking of the Civilized World* (New York, Boston, and London: Back Bay Books, 2017), pp. 146-148.

[20] City of Boston, "Preparing for Climate Change," boston.gov/departments/environment/climate-ready-boston #neighborhoods, last updated September 10, 2019, retrieved October 6, 2019.

[21] Goodell, pp. 238-240.

[22] Ibid., pp. 121-134. The impact on coastal wetlands was not anticipated. Erik Stokstad, "Venice's barrier against rising seas could jeopardize city's ecosystem: starved of sediment, salt marshes may not be able to keep up with rising sea level," *Nature,* November 29, 2021.

[23] Ibid. Goodell, pp. 190-195. The problem may be unsolvable, however. Although the base itself may be saved, the roads leading to it, along which all the sailors will need to commute each day, will be subject to the flooding.

[24] Ibid, pp. 106-110.

[25] Cement production, by the way, contributes to CO_2 emissions. IPCCAR5, p. 12.

[26] It is sad to think that the American West, with its imaginative pull, its inspiration to artists, and its vital rejuvenating energy, will lose its character. On the other hand, what may emerge from the change are population centers that may be appropriately sustained (unlike Phoenix and Las Vegas) and wilderness elsewhere.

[27] C. Turley, T. Kelzer, P. Williamson, J.-P. Gattuso, P. Ziverl, R. Monroe, K. Boot, M. Huelsenbeck, "Hot, Sour and Breathless—Ocean under stress," 2016. Partners: Plymouth Marine Laboratory, UK Ocean Acidification Research Programme, European Project on Ocean Acidification, Mediterranean Sea Acidification in a Changing Climate Project, Scripps Institution of Oceanography at UC San Diego, Oceana. Available at public.wmo.int/en/resources/bulletin/hot-sour-and-breathless--ocean-under-stress, retrieved October 6, 2019.

[28] Please, might we have a better word than desertification? It was hard enough to write acidification just now. If Robert McNamara had come up with a better term than Vietnamization (perhaps *Countries are not dominoes)*, American confidence and tolerance of others might be very different now.

[29] In London in 1952 twelve thousand people died during a three-day smog. In Northern China over recent decades, thousands died from burning the free coal provided them by their government. In Beijing and Delhi and other cities around the world, including the United States (wise giver of the Clean Air Act), millions of lives are being cut short each year. Jonathan Mingle, "Our Lethal Air," New York Review of Books, September 26, 2019, pp. 64-68. Wallace-Wells, p. 103.

[30] Sciencedaily.com, "Failing phytoplankton, failing oxygen: Global warming disaster could suffocate life on planet Earth," December 1, 2015. Cited by McKibben, p. 34.

[31] McKibben, p. 35.

[32] BBC Two, "The Big Chill," November 13, 2003, bbc.co.uk/science/horizon/2003/bigchill.shtml, now archived, retrieved October 4, 2019. This theory is controversial. See Richard Seager, "Climate mythology: The Gulf Stream, European climate, and abrupt change," ocp.ldeo.columbia.edu/res/div/ocp/gs, retrieved October 4, 2019.

[33] Wallace-Wells, p. 110.

[34] McKibben, pp. 34-35.

Notes

[35] Nick Bostrom, "Existential Risks: Analyzing Human Extinction Scenarios and Related Hazards," Journal of Evolution and Technology, 2002, v. 9 p.7. Cited in Wallace-Wells, p. 173.

[36] Lest, as I said in the Preface, we squander our limited resources. With real threats looming, it is incredibly frustrating what we humans have chosen to fight about. I suppose that's our character, still to evolve. Or perhaps the sign of a hopeful vitality, that such conflicts are possible in a civilization so beset. But all this grasping for advantage seems under the circumstances like the activities not of the great race we aspire to be but of "nervous little hopper men," as Willa Cather once described them ("Two Friends" in *Obscure Destinies* (New York: Alfred A. Knopf, 1932), p. 197), unable to see the world at large.

[37] CLOCK, pp. 2ff. In their 2019 issue the atomic scientists chose as an aggravating factor—perhaps as a threat itself, the writing isn't entirely clear—the "rise during the last year in the intentional corruption of the information ecosystem on which modern civilization depends"—in other words, disinformation, organized lying.

[38] Ibid., pp. 18, 20. Recall that 1991 was three years after James Hansen's warning about climate change. Thus has the climate problem stolen upon our consciousness, first on cat feet and then by a tiger's leap.

[39] Wikipedia, "Treaty on the Non-Proliferation of Nuclear Weapons," September 25, 2019, retrieved September 28, 2019.

[40] Wikipedia, "List of states with nuclear weapons," October 17, 2019, retrieved October 17, 2019.

[41] William J. Perry and Tom Z. Collina, *The Button: the new nuclear arms race and presidential power from Truman to Trump* (Dallas: BenBella Books, 2020), pp. 199-200.

[42] It has not formally withdrawn from the treaty, but inspections of sites, a vital component of compliance, have been stopped. Wikipedia, "New START," August 15, 2023, retrieved August 15, 2023.

[43] Wikipedia, "Intermediate-Range Nuclear Forces Treaty," October 2, 2019, retrieved October 16, 2019.

[44] Hans M. Kristensen and Matt Korda, "Tactical nuclear weapons, 2019," Bulletin of the Atomic Scientists: Nuclear Notebook, 2019, 75:5, 252-261, DOI: 10.1080/00963402 .2019.1654273.

[45] Wikipedia, "Asteroid impact avoidance," October 11, 2019, retrieved October 17, 2019.

[46] Jonathan Swan, Margaret Talev, "Scoop: Trump suggested nuking hurricanes to stop them from hitting U.S.," Axios, August 25, 2019, axios.com/trump-nuclear-bombs-hurricanes-97231f38-2394-4120-a3fa-8c9cf0e3f51c.html, retrieved October 7, 2019.

[47] Malcolm Gladwell, *The Tipping Point* (Boston: Little, Brown and Company, 2000), p. 219.

[48] See Steven Spielberg, director, *Indiana Jones and the Kingdom of the Crystal Skull,* Lucasfilm, distributed by

Paramount Pictures, 2008, in which Indy, finding himself at the Nevada Test Site, a model of 1957 suburbia, takes shelter in a lead-lined refrigerator just before the blast. The refrigerator is sent skyward like Dorothy's house in the tornado, but Indy survives, of course, and after a cursory decontamination by the FBI, apparently trained in the practice, carries on with his adventures. This sequence is not alone in dramatic film, even those purporting to show the enormity of nuclear weapons.

[49] Life Magazine, April 19, 1954, books.google.com /books?id=N1MEAAAAMBAJ, retrieved October 3, 2019.

[50] Wikipedia "Multiple independently targetable reentry vehicle," September 7, 2019, retrieved October 4, 2019.

[51] WARN1.

[52] WARN2.

[53] U.N. Development Programme, "Human Development Indices and Indicators: 2015 Statistical Update," summarized in un.org/sustainabledevelopment/blog/2015/12/2-billion-move-out-of-extreme-poverty-over-25-years-says-un-report, retrieved October 4, 2019. Note that the link in the blog now accesses the 2018 report.

[54] Ibid.

[55] Wikipedia, "Tragedy of the commons," September 13, 2019, retrieved September 16, 2019.

[56] Wikipedia, "Garrett Hardin," October 16, 2019, retrieved November 8, 2019.

[57] Goodell, p. 173.

[58] Ibid., p. 176. This means, I take it, that forty-six million gallons of brine must be returned to the ocean without appreciably raising its salinity.

[59] A difficult problem. If the iceberg is too small, the trip wouldn't be worth it; if too large, it would be too heavy or cumbersome to tow. One design has the "tug" lassoing the iceberg with a giant ribbon. Nypost.com/2017/05/17/dubai -wants-to-drag-icebergs-from-antarctica-for-freshwater, retrieved October 4, 2019. Independent.co.uk/news/world/ middle-east/uae-icebergs-drinking-water-from-antarctica- towed-united-arab-emirates-a7715561.html, retrieved October 4, 2019.

[60] Wikipedia, "Lithium," October 11, 2019, retrieved October 18, 2019.

[61] Wikipedia, "1973 oil crisis," September 10, 2019, retrieved October 19, 2019. The difference between the crude and retail prices can be explained by several factors, including price controls, Federal quotas imposed upon the states, and withdrawal of oil from the strategic reserve. There was still enough oil. There was not the perception of enough oil.

[62] Ibid.

[63] Wikipedia, graph of "U.S. Crude Oil Production and Imports," upload.wikimedia.org/wikipedia/commons/a/a0/US_ Crude_Oil_Production_and_Imports.svg, retrieved October 19, 2019.

[64] The role of the U.N., as we've seen, is moral and philanthropic; members follow its policies as they wish. The same limitations apply to international law, which is no more powerful or enduring than the treaty that created it. One of the lesser reported crimes against humanity of Russia's invasion of Ukraine is that it has diverted resources and attention and fractured cohesion when all our efforts are needed to deal with the overheated planet.

[65] When we think of transformative technologies today, we think of the internet, personal computers, smartphones, television, and AI. But not so long ago, if you asked a housewife in rural America what had transformed her life, she would say electricity and her washing machine.

[66] Depending on a software system without knowing when or how to overrule it were common factors in four recent transportation disasters: the crashes of the Boeing 747 MAX in 2018 and 2019 and the collisions of the Navy destroyers Fitzgerald and John S. McCain with merchant ships in 2017.

[67] Henry A. Kissinger, et. al., *The Age of AI and Our Human Future* (New York: Back Bay Books, 2021), pp. 135-176.

[68] I am reminded of this whenever a filmmaker uses a POV camera unrealistically. You know the technique: the camera follows a character, usually walking, as if we the audience are seeing with his eyes. The mistake is to give that shot a disjointed, uneven feel, for the reality is that no matter how bumpy the terrain or how awkwardly the character moves, what we see is smoothed out by the world's finest gyro stabilizer, the human brain.

[69] Organization for Economic Cooperation and Development, "Working Age Population," 2018, data.oecd.org/pop/working-age-population.htm, retrieved October 20, 2019.

[70] IPCC1.5, pp. 4ff.

[71] Meaning that climate feedbacks would accelerate warming and prevent mitigation. Umair Urfan, "A major new climate report slams the door on wishful thinking," Vox, 10/7/2018, vox.com/2018/10/5/17934174/climate-change-global-warming-un-ipcc-report-1-5-degrees, retrieved 10/4/2019.

[72] WARN2, p. 1.

[73] It could have meant, for example, intellectual, spiritual, or behavioral improvement; greater knowledge or accomplishment.

[74] If there was ever a time for using the superlative degree with the word *perfect,* it is this.

[75] Wikipedia, "Lay's", October 3, 2019, retrieved October 8, 2019.

[76] In 2022 the adult obesity prevalence in the United States was 41.9 percent. Centers for Disease Control and Prevention, "Adult Obesity Facts," https://www.cdc.gov/obesity/data/adult.html, May 17, 2022, retrieved August 18, 2023.

[77] Centers for Disease Control and Prevention, "Adult Obesity Causes & Consequences," cdc.gov/obesity/adult/causes.html, August 29, 2017, retrieved October 8, 2019. Nothing in this discussion is meant to dismiss the other contributions to this highly complicated epidemic: food

insufficiency, food deserts, economic inequality, sedentary lifestyles, other physical diseases, mental health issues, and related types of addiction. But, I believe, the most important contribution and the way this problem started in the first place lie within the basic arithmetic: calories consumed versus calories expended.

[78] Or started. We still don't know the long-term effects of consuming marijuana. CBD, which contains lower levels of THC, is having a boom although its efficacy has been proved for only a few disorders.

[79] Wikipedia, "United States presidential approval rating," Gallup Poll Approval Rating for Jimmy Carter, September 27, 2019, retrieved September 28, 2019.

[80] Wikipedia, "Moral Equivalent of War speech," September 22 2019, retrieved September 26, 2019.

[81] Ibid.

[82] Nor, many would add, his brilliant career as a humanist and writer after he left office. In effect Carter as conservation advocate was trying to solve a problem no one cared about—today's situation turned inside-out.

[83] Quoted in Goodell, p. 212.

[84] Stephen Leary, "We have too many fossil-fuel power plants to meet climate goals," National Geographic, nationalgeographic.com/environment/2019/07/we-have-too-many-fossil-fuel-power-plants-to-meet-climate-goals, July 1, 2019, retrieved October 9, 2019.

[85] In an address to the United Nations Climate Action Summit in September 2019, sixteen-year-old Greta Thunberg of Sweden cried, "This is all wrong. I shouldn't be standing here. I should be back in school on the other side of the ocean. Yet you all come to me for hope? How dare you! You have stolen my dreams and my childhood with your empty words. And yet I'm one of the lucky ones. People are suffering. People are dying. Entire ecosystems are collapsing. We are in the beginning of a mass extinction. And all you can talk about is money and fairytales of eternal economic growth. How dare you...! You are failing us.... But the young people are starting to understand your betrayal. The eyes of all future generations are upon you. And if you choose to fail us, I say: We will never forgive you." Wikipedia, "Greta Thunberg," September 28, 2019, retrieved September 28, 2019.

[86] Wikipedia, "National Maximum Speed Law," September 22, 2019, retrieved October 9, 2019.

[87] Wendy's introduced the question as an advertising slogan in 1984. Walter Mondale, the Democratic candidate for president that year, picked it up in his campaign. Wikipedia, "Where's the beef?", August 8, 2019, retrieved September 22, 2019. Methane, produced by cattle and by leaks in natural gas production and distribution, occupies a smaller proportion of greenhouse gases than carbon dioxide but is many times more harmful in trapping sunlight and warming the atmosphere.

[88] Transportation—29%, Electricity—28%, Industry—22%, Commercial and Residential—12%, Agriculture—9%. United States Environmental Protection Agency, "Sources of Greenhouse Gas Emissions," epa.gov/ghgemissions/sources-greenhouse-gas-emissions, retrieved October 24, 2019.

[89] Wikipedia, "Highway Trust Fund," October 27, 2017, retrieved October 26, 2019.

[90] Americans used to relish a challenge—other countries admired that about us—but we gave up too easily converting to metric, and it has hurt us ever since; both the stubborn adherence to an outmoded, exceptional (not in a good way), and inefficient standard of measure and the giving up. Metric isn't that hard. We should try again.

[91] Why? First, because there won't be enough money to continue the subsidies. A policy almost ninety years old and started in a time of national emergency has come to feel like a way of life, but as incomes fall and claims for government assistance rise, the American people will be ever more inclined to adopt a Spencerian view toward the fate of their fellow citizens, particularly when that fate results from their own choices. And second, because farmers will need to experiment, taking their chances on success or failure. Government is cautious from ignorance, experience, group-think, and fear of adverse consequences. U.S. farm policy is the classic dromedary designed by committee. If the first farms had depended on government assistance and followed government rules, humankind would still be hunter-gatherers.

[92] Wikipedia, "National debt of the United States," August 8, 2023, retrieved August 15, 2023. A headline in the Washington Post on October 25, 2019 reports that the U.S. budget deficit—that is, the annual Federal deficit—added 984 billion to the national debt in 2019, a surprisingly high figure during "relatively good economic times." This, of course, was before the added expenditures resulting from the Covid-19 pandemic

[93] Trading Economics, "United States Private Debt to GDP," 2019, tradingeconomics.com/united-states/private-debt-to-gdp, retrieved October 25, 2019.

[94] When One World Trade Center was built after the twin towers collapsed in the 9/11 attacks, it was a point of pride that new construction methods made it strong enough to withstand an airliner.

[95] Wikipedia, "Infrastructure and Investment Jobs Act," August 14, 2023, retrieved August 16, 2023.

[96] As an early step, disaster tourism needs to stop: if people won't stop coming to gape at glaciers and the companies won't refuse to carry them, the governments concerned should prohibit the practice. There is no right to use an aircraft just to confirm what other people have seen. By now, when the relevant stories have been reported more than once, this goes for journalists too.

[97] More Than Shipping, "The Carbon Footprint of the Shipping Industry," 2018, morethanshipping.com/green-shipping, retrieved October 25, 2019.

[98] Wikipedia, "Maglev," October 25, 2019, retrieved October 26, 2019.

[99] Wikipedia, "Vactrain," October 11, 2019, retrieved October 26, 2019.

[100] For an inspiring look at what has already been done in Milan and other Italian cities, see the work of architect Luciano Pia (lucianopia.it, retrieved September 23, 2019).

[101] "Fore'handedness n. (US) the action of looking to the future; prudence, foresight." Leslie Brown, ed., *The New Shorter Oxford English Dictionary* (Oxford: Clarendon Press, 1993).

[102] Adding two bytes to the memory space for the century was easy enough to do, but the bug was so prevalent—in trillions of lines of code—that an enormous effort was required around the world to fix it. Perhaps this should give us hope that every little bit that hurts can be fixed as well, before the next perceived Apocalypse.

[103] Alan Stern and David Grinspoon, *Chasing New Horizons: Inside the epic first mission to Pluto* (New York: Picador, 2018), pp. 5, 112-113. As it passed Pluto, on its way to the Kuiper Belt, New Horizons was traveling at 33,000 mph, according to the authors. Seventeen years after leaving Earth, it is still outbound.

[104] Goodell, p. 56.

[105] Wallace-Wells, p. 170.

[106] Wikipedia, "Carbon capture and storage: CO_2 transportation," October 9, 2019, retrieved October 11, 2019.

[107] Wallace-Wells, p. 107.

[108] Goodell, p. 262.

[109] *Plus ça change, plus c'est la même chose*—"The more things change, the more they remain the same." Jean-Baptiste Alphonse Karr, 1848, cited in Wikipedia,

"Jean-Baptiste Alphonse Karr," September 11, 2019, retrieved October 27, 2019.

[110] "[T]o form a more perfect Union, establish Justice, insure domestic Tranquility, provide for the common defence, promote the general Welfare, and secure the Blessings of Liberty to ourselves and our Posterity...."

[111] Here is just a partial list of important problems since the moon landings our country has failed to solve:

- War powers of the Executive and proper oversight by the Congress
- The War on Drugs
- Converting to the metric system
- Gun violence
- Finding sensible, moral solutions to unwanted pregnancy and the uses of human embryos
- The gender gap in incomes and other discrimination
- The sharp rise in the suicide rate
- Immigration and undocumented residents
- The rise of the prison population, predominantly among African-Americans
- The projected Social Security shortfall
- The projected Medicare shortfall
- The poor quality of public K-12 education
- The high cost of post-secondary education
- The high cost of healthcare and inequitable access
- Budget deficit and the national debt
- The prevalence of short-term funding
- The periodic crisis over the debt ceiling
- The abuses of the internet and social networking
- The normalization of pornography
- The obesity epidemic

Notes

¹¹² Not quite. I am leaving out the distinction drawn by the late Justice Scalia between strict construction and textualism; in describing the two sides in the larger conflict, it makes no difference.

¹¹³ The sentence begins with a present participle absolute construction, an archaism of grammar seldom used or understood today. The introductory clause is causal (e.g., *"Because* a well regulated Militia is necessary to the security of a free State, the right of the people to keep and bear arms shall not be infringed"). The assertion isn't true any longer—none but a few self-appointed militiamen would say that it was—so it may not logically establish the main clause. Nor, for that matter, does its obsolescence disestablish the main clause. We must decide for our own time how to bear arms and how many and what kind and with what precautions.

¹¹⁴ The 27ᵗʰ Amendment, which prohibits each Congress from raising its own pay, hardly counts, having been submitted in 1789 but not approved until 1992, a kind of cleaning out of the attic. The 26ᵗʰ Amendment was approved in 1971. The Equal Rights Amendment was ratified by Virginia in January 2020, reaching the required three-quarters threshold, but Congress when sending it to the states in 1972 had established a time limit, which, although extended once, has since expired. The status of the amendment is therefore unclear and may only be determined by litigation. With no other amendment working through the states, at least fifty years will have passed since the last change. Wikipedia, "List of amendments to the United States Constitution," October 24, 2019, retrieved October 28, 2019. ABC News, "Equal Rights Amendment ratified in Virginia, reaching

historic national threshold," January 15, 2020, retrieved January 31, 2020.

[115] This was written before January 6, 2021.

[116] I may be accused here of self-contradiction in describing the perfected process that I criticized as time-wasting in the passage about sea barriers. But that was at the local level, where the problem was clear and the solution largely technical; here the problem is to build support for a long-term policy requiring great sacrifice, which must be done right the first time.

[117] "[S]ociologist, historian, civil rights activist, Pan-Africanist, author, writer, and editor," as Wikipedia puts it ("W.E.B. Du Bois," October 7, 2019, retrieved October 21, 2019), he effectively assumed the role of the nation's conscience on matters of race relations after the death of Frederick Douglass. With others he founded the National Association for the Advancement of Colored People. He died a year before the signing of the Civil Rights Act of 1964.

[118] Quoted in Taylor Branch, *Parting the Waters: America in the King Years 1954-63* (New York: Simon and Schuster, 1988), P. 563.

[119] Yang2020.com/policies/the-freedom-dividend, retrieved September 22, 2019.

[120] Lindsay M. Howden and Julie A. Meyer. U.S. Department of Commerce, U.S. Census Bureau. "Age and Sex Composition: 2010." Washington: May 2011.

¹²¹ Wikipedia, "2019 U.S. federal budget," August 2, 2019, retrieved September 22, 2019.

¹²² Opportunity.businessroundtable.org/wp‑content/ uploads/2019/09/BRT‑Statement‑on‑the‑Purpose‑of‑ a ‑corporation‑with‑signatures‑1.pdf, retrieved October 6, 2019.

¹²³ And shorter lives. See Nicholas Kristof on loneliness and life span: "One of the Best Things We Can Do for Our Health is Free—and Fun," *The New York Times,* September 6, 2023.

¹²⁴ Followed by fifty‑five years of economic disadvantage with no improvement in sight.

¹²⁵ McKibben (p. 202) says, "[A]round the world, polling shows that people are not just highly concerned about global warning, but also willing to pay a price to solve it. Americans, for instance, said in 2017 that they were willing to see their energy bills rise 15 percent and have the money spent on clean energy programs...." The real sacrifice will be much greater than 15 percent, and people will have to keep paying it long after they are tired of sacrificing.

¹²⁶ Expressed, these days, in angry denunciation.

¹²⁷ What is instant replay but a product of the desire for justice, the love of precision, and the fear of criticism?

¹²⁸ McKibben, p. 176.

¹²⁹ I have always felt smug about this achievement, which occurred while I was the navigator of a destroyer

escort in 1973, until I read that the New Horizons spacecraft traveled three billions miles over more than nine years to reach a closest point of approach to Pluto within seven minutes of its scheduled time. Stern and Grinspoon, pp. 241-244.

[130] Air Transport Action Group, "CORSIA Explained," aviationbenefits.org/environmental-efficiency/climate-action/offsetting-emissions-corsia/corsia, retrieved October 10, 2019.

[131] Center for Climate and Energy Solutions, "Reducing Your Transportation Footprint," c2es.org/content /reducing-your-transportation-footprint, retrieved October 10, 2019.

[132] If enough of us can wrest the housing stock from investors in the sub-lease market.

[133] Yian Q. Mui, *Washington Post,* "The shocking number of Americans who can't cover a $400 expense," May 25, 2016.

[134] Emily Butdorf, "Key Savings Statistics and Trends in 2023," *Forbes Advisor,* June 29, 2023, retrieved August 19, 2023.

[135] In the extended fire season of October 2019, California shut off the power to thousands of customers out of fear of wildfires caused by windblown power lines. At one stroke Pacific Gas & Electric made a generation of pioneers. Hannah Fry, Alejandra Reyes-Velarde, Richard Winton, "Edison begins power shutdowns in Southern California amid winds, fire danger," latimes.com, retrieved October 10, 2019.

Notes

[136] In 2014 the top one percent of Americans possessed forty percent of the country's wealth; in 2017 the eight richest people in the world, six of them Americans, possessed as much wealth as the bottom half of the world's population. Wikipedia, "Wealth inequality in the United States," October 8, 2019, retrieved October 11, 2019.

[137] Pamela Falk, "Climate negotiators strike deal to slow global warming," CBS News, archived from the original on December 13 2015. Cited in Wikipedia, "Paris Agreement," August 1, 2023, retrieved August 16, 2023.

[138] The information in this passage about the United Nations Framework Convention on Climate Change, of which the Paris Agreement is one instrument, was drawn primarily from Wikipedia, "Paris Agreement," October 7, 2019, retrieved October 12, 2019. The explanation of "cap-and-trade" and the skepticism about the potential success of adaptation and loss and damage measures are my own.

[139] Oxford: Oxford University Press, pp. 3-10.

[140] "Displacement of populations and destruction of cultural language and tradition is equivalent in our minds to genocide." —Tony de Brum, Foreign Minister of the Marshall Islands, in a radio interview on October 6, 2015, reported by Goodell, p. 170.

[141] Green Climate Fund, *Pledge Tracker,* July 31, 2023, retrieved August 16, 2023. Wikipedia, "Green Climate Fund," September 28, 2019, retrieved October 12, 2019.

[142] UNEP18, p. 8.

[143] If any of the suppliers themselves proved hypocritical on the subject, as Australia and China are, penalties could be applied to them. UNEP18, p. 8.

[144] Wikipedia, "List of countries by distribution of wealth," October 6, 2019, retrieved October 13, 2019.

[145] Wikipedia, "List of countries by carbon dioxide emissions," October 11, 2019, retrieved October 13, 2019.

[146] My calculation: 330 million Americans divided by 8 billion people in the world.

[147] A U.N. projection cited by Wallace-Wells, p. 7.

[148] My calculation: 330 million Americans divided by 3.8 million square miles (=87) increased to 358 million Americans divided by 3.8 million square miles (=95). (Results have been rounded up to the nearest person.) This assumes that the habitable land area of the U.S. remains the same, which is unlikely in view of the rising seas, drought, and wildfire. If half the area of the U.S. were lost—an extremely high assumption—the population density would increase to 189 per square mile, which would move the U.S. from 146th to 107th in the world, with about the same density as Uzbekistan and Brunei.

www.ingramcontent.com/pod-product-compliance
Lightning Source LLC
Chambersburg PA
CBHW070105070426
42448CB00038B/1668